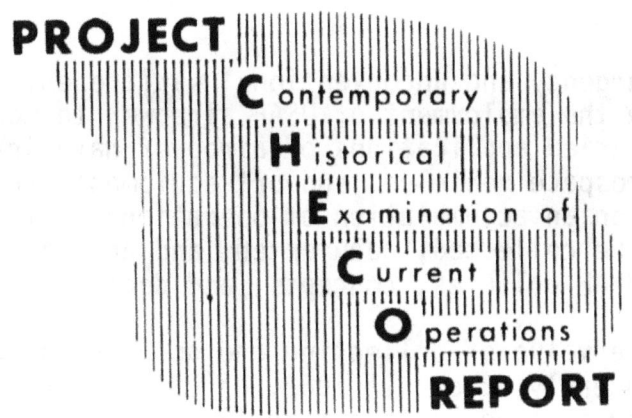

PROJECT RED HORSE

1 SEPTEMBER 1969

HQ PACAF

Directorate, Tactical Evaluation

CHECO Division

Prepared by:
LT DEREK H. WILLARD
Project CHECO 7th AF, DOAC

PROJECT CHECO REPORTS

The counterinsurgency and unconventional warfare environment of Southeast Asia has resulted in the employment of USAF airpower to meet a multitude of requirements. The varied applications of airpower have involved the full spectrum of USAF aerospace vehicles, support equipment, and manpower. As a result, there has been an accumulation of operational data and experiences that, as a priority, must be collected, documented, and analyzed as to current and future impact upon USAF policies, concepts, and doctrine.

Fortunately, the value of collecting and documenting our SEA experiences was recognized at an early date. In 1962, Hq USAF directed CINCPACAF to establish an activity that would be primarily responsive to Air Staff requirements and direction, and would provide timely and analytical studies of USAF combat operations in SEA.

Project CHECO, an acronym for Contemporary Historical Examination of Current Operations, was established to meet this Air Staff requirement. Managed by Hq PACAF, with elements at Hq 7AF and 7AF/13AF, Project CHECO provides a scholarly, "on-going" historical examination, documentation, and reporting on USAF policies, concepts, and doctrine in PACOM. This CHECO report is part of the overall documentation and examination which is being accomplished. Along with the other CHECO publications, this is an authentic source for an assessment of the effectiveness of USAF airpower in PACOM.

MILTON B. ADAMS, Major General, USAF
Chief of Staff

DISTRIBUTION LIST

1. SECRETARY OF THE AIR FORCE

    a. SAFAA 1
    b. SAFLL 1
    c. SAFOI 2

2. HEADQUARTERS USAF

    a. AFBSA 1

    b. AFCCS
        (1) AFCCSSA 1
        (2) AFCVC 1
        (3) AFCAV 1
        (4) AFCHO 2

    c. AFCSA
        (1) AFCSAG 1
        (2) AFCSAMI 1

    d. AFGOA 2

    e. AFIGO
        (1) AFISI 3
        (2) AFISP 1

    f. AFMSG 1

    g. AFNIN
        (1) AFNIE 1
        (2) AFNINA 1
        (3) AFNINCC 1
        (4) AFNINED 4

    h. AFAAC 1
        (1) AFAMAI 1

    i. AFODC
        (1) AFOAP 1
        (2) AFOAPS 1
        (3) AFOCC 1
        (4) AFOCE 1
        (5) AFOMO 1

    j. AFPDC
        (1) AFPDPSS 1
        (2) AFPMDG 1
        (3) AFPDW 1

    k. AFRDC 1
        (1) AFRDD 1
        (2) AFRDQ 1
        (3) AFRDQRC 1
        (4) AFRDR 1

    l. AFSDC
        (1) AFSLP 1
        (2) AFSME 1
        (3) AFSMS 1
        (4) AFSPD 1
        (5) AFSSS 1
        (6) AFSTP 1

    m. AFTAC 1

    n. AFXDC
        (1) AFXDO 1
        (2) AFXDOC 1
        (3) AFXDOD 1
        (4) AFXDOL 1
        (5) AFXOP 1
        (6) AFXOSL 1
        (7) AFXOSN 1
        (8) AFXOSO 1
        (9) AFXOSS 1
        (10) AFXOSV 1
        (11) AFXOTR 1
        (12) AFXOTW 1
        (13) AFXOTZ 1
        (14) AFXOXY 1
        (15) AFXPD 6
            (a) AFXPPGS 3

3. MAJOR COMMANDS

   a. TAC

      (1) HEADQUARTERS
         (a) DO 1
         (b) DPL 2
         (c) DOCC. 1
         (d) DORQ. 1
         (e) DIO 1

      (2) AIR FORCES
         (a) 12AF
             1. DORF 1
             2. DI 1
         (b) 19AF(DI) 1
         (c) USAFSOF(DO) 1

      (3) WINGS
         (a) 1SOW(DO) 1
         (b) 4TFW(DO) 1
         (c) 23TFW(DOI) 1
         (d) 27TFW(DOI) 1
         (e) 33TFW(DOI) 1
         (f) 64TAW(DOI) 1
         (g) 67TRW(C) 1
         (h) 75TRW(DO) 1
         (i) 316TAW(DOP) 1
         (j) 317TAW(EX) 1
         (k) 363TRW(DOC) 1
         (l) 464TAW(DO) 1
         (m) 474TFW(TFOX) 1
         (n) 479TFW(DOF) 1
         (o) 516TAW(DOPL) 1
         (p) 4410CCTW(DOTR) . . . 1
         (q) 4510CCTW(DO16-I) . . 1
         (r) 4554CCTW(DOI) . . . 1

      (4) TAC CENTERS, SCHOOLS
         (a) USAFTAWC(DA) 2
         (b) USAFTARC(DID) . . . 2
         (c) USAFTALC(DCRL) . . . 1
         (d) USAFTFWC(CRCD) . . . 1
         (e) USAFAGOS(DAB-C) . . 1

   b. SAC

      (1) HEADQUARTERS
         (a) DOPL. 1
         (b) DPLF. 1
         (c) DM. 1
         (d) DI. 1
         (e) OA. 1
         (f) HI. 1

      (2) AIR FORCES
         (a) 2AF(DICS) 1
         (b) 15AF(DI) 1

      (3) AIR DIVISIONS
         (a) 3AD(DO) 3

   c. MAC

      (1) HEADQUARTERS
         (a) MAOID 1
         (b) MAOCO 1
         (c) MACHO 1
         (d) MACOA 1

      (2) AIR FORCES
         (a) 21AF(OCXI) 1
         (b) 22AF(OCXI) 1

      (3) WINGS
         (a) 61MAWg(OIN) 1
         (b) 62MAWg(OCXP) . . . 1
         (c) 436MAWg(OCXC) . . . 1
         (d) 437MAWg(OCXI) . . . 1
         (e) 438MAWg(OCXC) . . . 1

      (4) MAC SERVICES
         (a) AWS(AWXW) 1
         (b) ARRS(ARXLR) 1
         (c) ACGS(AGOV) 1
         (d) AAVS(AVODOD) . . . 1

d. ADC

   (1) HEADQUARTERS
      (a) ADODC 1
      (b) ADOOP 1
      (c) ADLCC 1

   (2) AIR FORCES
      (a) AF ICELAND(FICAS) . . 2

   (3) AIR DIVISIONS
      (a) 25AD(ODC) 2
      (b) 29AD(ODC) 1
      (c) 33AD(OIN) 1
      (d) 35AD(CCR) 1
      (e) 37AD(ODC) 1

e. ATC

   (1) HEADQUARTERS
      (a) ATXPP 1

f. AFLC

   (1) HEADQUARTERS
      (a) MCVSS 1
      (b) MCNAP 1

g. AFSC

   (1) HEADQUARTERS
      (a) SCLAP 3
      (b) SCS-6 1
      (c) SCGCH 2
      (d) SCTPL 1
      (e) ASD(ASJT) 1
      (f) ESD(ESO) 1
      (g) RADC(EMOEL) 2
      (h) ADTC(ADGT) 1

h. USAFSS

   (1) HEADQUARTERS
      (a) ODC 1
      (b) CHO 1

   (2) SUBORDINATE UNITS
      (a) Eur Scty Rgn(OPD-P) . 1
      (b) 6940 Scty Wg(OOD) . . 1

i. AAC

   (1) HEADQUARTERS
      (a) ALDOC-A 2

j. USAFSO
   (1) HEADQUARTERS
      (a) COH 1

k. PACAF

   (1) HEADQUARTERS
      (a) DP 1
      (b) DI 1
      (c) DPL 2
      (d) CSH 1
      (e) DOTEC 5
      (f) DE 1
      (g) DM 1
      (h) DOTECH 1

   (2) AIR FORCES
      (a) 5AF(DOPP) 1
      (b) Det 8, ASD(DOASD) . . 1
      (c) 7AF
         1. DO 1
         2. DIXA 1
         3. DPL 1
         4. TACC 1
         5. DOAC 2
      (d) 13AF
         1. CSH 1
         2. DPL 1
      (e) 7/13AF(CHECO) 1

   (3) AIR DIVISIONS
      (a) 313AD(DOI) 1
      (b) 314AD(DOP) 2
      (c) 327AD
         1. DO 1
         2. DI 1
      (d) 834AD(DO) 2

(4) WINGS
    (a) 8TFW(DCOA). 1
    (b) 12TFW(DCOI) 1
    (c) 35TFW(DCOI) 1
    (d) 37TFW(DCOI) 1
    (e) 56SOW(WHD). 1
    (f) 347TFW(DCOOT) 1
    (g) 355TFW(DCOC). 1
    (h) 366TFW(DCO) 1
    (i) 388TFW(DCO) 1
    (j) 405TFW(DCOA). 1
    (k) 432TRW(DCOI). 1
    (l) 460TRW(DCOI). 1
    (m) 475TFW(DCO) 1
    (n) 633SOW(DCOI). 1
    (o) 1st Test Sq(A). 1

(5) OTHER UNITS
    (a) Task Force ALPHA(DXI) . . . 1
    (b) 504TASG(DO) 1

m. USAFE

(1) HEADQUARTERS
    (a) ODC/OA. 1
    (b) ODC/OTA 1
    (c) OOT 1
    (d) XDC 1

(2) AIR FORCES
    (a) 3AF(ODC). 2
    (b) 16AF(ODC) 2
    (c) 17AF
        1. ODC. 1
        2. OID. 1

(3) WINGS
    (a) 20TFW(DCOI) 1
    (b) 36TFW(DCOID). 1
    (c) 50TFW(DCO). 1
    (d) 66TRW(DCOIN-T). 1
    (e) 81TFW(DCOI) 1
    (f) 401TFW(DCOI). 1
    (g) 513TAW(OID) 1
    (h) 7101ABW(DCO-CP) 1

4. SEPARATE OPERATING AGENCIES

  a. ACIC(ACOMC). 2
  b. AFRES(AFRXPL). 2
  c. USAFA
    (1) CMT. 1
    (2) DFH. 1
  d. AU
    (1) ACSC-SA. 1
    (2) AUL(SE)-69-108 2
    (3) ASI(ASD-1) 1
    (4) ASI(ASHAF-A) 2
  e. AFAFC(EXH) 1

5. MILITARY DEPARTMENTS, UNIFIED AND SPECIFIED COMMANDS, AND JOINT STAFFS

    a. COMUSJAPAN. 1
    b. CINCPAC 1
    c. COMUSKOREA. 1
    d. COMUSMACTHAI. 1
    e. COMUSMACV 1
    f. COMUSTDC. 1
    g. USCINCEUR 1
    h. USCINCSO. 1
    i. CINCLANT. 1
    j. CHIEF, NAVAL OPERATIONS 1
    k. COMMANDANT, MARINE CORPS. 1
    l. CINCONAD. 1
    m. DEPARTMENT OF THE ARMY. 1
    n. JOINT CHIEFS OF STAFF 1
    o. JSTPS 1
    p. SECRETARY OF DEFENSE (OASD/SA). 1
    q. USCINCMEAFSA. 1
    r. CINCSTRIKE. 1
    s. CINCAL. 1
    t. MAAG-China/AF Section (MGAF-O). 1
    u. Hq Allied Forces Northern Europe (U.S. Documents Office). . . . 1

6. SCHOOLS

    a. Senior USAF Representative, National War College. 1
    b. Senior USAF Representative, Armed Forces Staff College. 1
    c. Senior USAF Rep, Industrial College of the Armed Forces 1
    d. Senior USAF Representative, Naval Amphibious School 1
    e. Senior USAF Rep, US Marine Corps Education Center 1
    f. Senior USAF Representative, US Naval War College. 1
    g. Senior USAF Representative, US Army War College 1
    h. Senior USAF Rep, US Army C&G Staff College. 1
    i. Senior USAF Representative, US Army Infantry School 1
    j. Senior USAF Rep, US Army JFK Center for Special Warfare 1
    k. Senior USAF Representative, US Army Field Artillery School. . . 1

TABLE OF CONTENTS

	Page
FOREWORD	x
CHAPTER I - INTRODUCTION	1
Mission	1
Command and Control	2
Organization	3
CHAPTER II - RED HORSE ORGANIZATIONS IN SOUTHEAST ASIA	5
Introduction	5
Phan Rang - 554th CES (HR)	5
555th CES (HR) Cam Ranh Bay	9
819th CES (HR) - Phu Cat	12
Construction Summary	15
820th CES (HR) Tuy Hoa/Da Nang	16
823d CES (HR) Bien Hoa	26
1st Civil Engineering Group	30
556th CES (HR) U-Tapao	32
CHAPTER III - TRAINING OF MILITARY PERSONNEL	37
CHAPTER IV - PROBLEMS AND SECOND THOUGHTS	40
CHAPTER V - CONCLUSION	46
FOOTNOTES	
Foreword	51
Chapter I	51
Chapter II	51
Chapter III	55
Chapter IV	55
APPENDIXES	
I. Projects Completed from 1 Sep 68 to 20 May 69	55
II. RED HORSE Combat Defense Teams	59
III. 555th CES (HR) Projects	62
IV. 820th CES Construction Program	70
V. 556th CES (HR) Project Summary	75
VI. Summary of Project CONCRETE SKY	94
GLOSSARY	97

FIGURES   Follows Page

1. Organization for Construction in SEA 2
2. Organization Chart, CES (HR) 4
3. RED HORSE Squadron/Detached Units in SEA 6
4. 554th RED HORSE Combat Defense 6
5. RED HORSE Concrete Batching Plant 6
6. Pouring Concrete Lane, Dispersal Hardstands 6
7. RED HORSE Defense Line 8
8. Forming Aircraft Shelters for Concrete Cover 8
9. Covering Aircraft Shelters with Crane/Bucket Method 8
10. Covering Aircraft Shelters with Squeeze Crete Pump 8
11. Aircraft Shelters with Concrete Cover 8
12. Construct Airmen's Dorms 8
13. Construct Aircraft Barrier (BAK-12) 8
14. RED HORSE Asphalt Mixing Plant 8
15. Constructed Railroad Spur 10
16. POL Pipline 10
17. Revetments, F-4C 12
18. Rocket Damage to RED HORSE Compound, Jan 1969 16

FOREWORD

A study requested by the Secretary of Defense in 1965 showed that "when national interests are involved and tactical forces are deployed without a declaration of national emergency or war, a quick-reacting, heavy repair force, organic to the Air Force, is essential."[1] Between June and September 1965, a study group from the Directorate of Civil Engineering at Headquarters USAF had analyzed the problem and obtained Air Staff approval to form such a force.

On 23 September 1965, the Tactical Air Command (TAC) was given responsibility for organizing, training, procuring equipment and supplies, and administering the formation of the first two Red Horse Squadrons (the 554th and 555th Civil Engineering Heavy Repair Squadrons). By 18 October 1965, Hq TAC at Langley Field, Virginia, completed and distributed a comprehensive programming plan covering the objectives, timetable of actions, reporting procedures, staffing requirements, and the naming of primary and subordinate unit project officers. The mission and capabilities of the squadrons, their limitations, and material requirements were also recorded.[2]

Thus, in the fall of 1965, responding to the changing military and political situation in Southeast Asia and the projected need for a rapid increase of U.S. military forces in that part of the world, Project RED HORSE was initiated. The rapidity of planning, organizing, and executing which characterized these early beginnings, was to become a permanent part of RED HORSE activities.

CHAPTER I

INTRODUCTION

Mission

Project RED HORSE, an acronym for Rapid Engineer Deployable Heavy Operations Repair Squadron, Engineer, exists: "to provide a mobile Civil Engineering unit organic to the Air Force that is manned, trained, and equipped to perform heavy repairs and upgrade airfields and facilities (constructed by other agencies) to support weapons systems deployed to a theater of operations. It accomplishes this objective through the Civil Engineering Squadrons (Heavy and Repair) (CES HR)." These two terms, RED HORSE and CES HR, are used interchangeably and refer to virtually the same Civil Engineering activity.[1]

The Air Force objective was to provide a rapid response capability to augment Base Engineer forces in the event of heavy bomb damage or disasters, as well as accomplish major repairs or construction where contract capability was not readily available. These squadrons were deployed with their own equipment and supplies, and their own mess and field dispensary, including a physician and medical technicians. Since they are Air Force-controlled squadrons, they retained their own identity in the field.[2]

As Maj. Gen. Robert H. Curtin, Director of the Air Force Civil Engineering, stated in his message to the "Air Force Civil Engineering Magazine":[3]

> *"These squadrons are to provide a continuing on-site...*
> *capability to meet operational needs of the combat zone.*
> *These units are not intended to, or do they in fact,*
> *minimize our reliance on Army Engineer Battalions for*
> *initial expeditionary Air Field work when available.*

1

> *Rather, these units will fill an existing gap in the broad spectrum of Civil Engineering capabilities needed by the Air Force to its operational missions."*

RED HORSE squadrons were deployed with their own weapons and were considered combat troops as well as engineers. When assigned to a base, they could be held responsible for perimeter defense as well as their engineering mission.

Command and Control

As explained in AFM 26-2 (Chap 3, 15 Mar 66), RED HORSE, or CES HR units, were Air Force-controlled units constituted and activated by Headquarters USAF and assigned to a Major Command. A Major Command could further assign the unit, unless the directive assigning it to the command prohibited further assignment or directed one specific assignment. Figure 1 is an organizational chart which depicts lines of control for Civil Engineering, including RED HORSE, in Southeast Asia.[4]

The policy on use of RED HORSE forces is outlined in AFR 85-25. Normally, the Army provided troop construction support and materials to accomplish new construction and rehabilitation projects for the Air Force. Air Force and Army component commanders jointly determined requirements for such work under existing policies and procedures (AFR 88-12) and the guidance of the unified commander. When additional engineering forces were needed to provide facilities for Air Force units, pending assignment of other departmental construction capability, then RED HORSE squadrons could be used.[5]

ORGANIZATION FOR CONSTRUCTION IN SEA

- PACNAV FACENGCO
- SEANAV FACENGCO
  - OICC
    - RMK-BPJ
  - NAVCONST BDE
- NAVFORV
  - NAVCONST BDE
  - SPECIALITY CONTRACTORS
- MACV
  - DIR CONST
- 7AF
  - BCE
  - 1ST CEG
    - 554TH CES (HR)
    - 555TH CES (HR)
    - 819TH CES (HR)
    - 820TH CES (HR)
    - 823D CES (HR)
    - PRIME BEEF
  - TURNKEY
- DAF
- DA
- USARV
  - USAFCV
  - PA & E
- SPECIALITY CONTRACTORS

Since RED HORSE squadrons were designed to operate under emergency contingency, or war conditions they did not, and were not always required to adhere to the work control, cost, or real property accounting procedures in AFMs 85-1, 170-5, or 93-1. Rather, the Major Command to which a squadron was assigned had the responsibility for developing, approving, and funding work programs. This responsibility could be delegated to a subordinate command.[6]

The Major Command would normally assign specific missions to the RED HORSE Squadrons under its operational control but, again, this responsibility could be delegated to the Numbered Air Force level.

Although Headquarters USAF had the responsibility for identifying requirements for the number of RED HORSE Squadrons and how these squadrons were to be tailored to meet the requirements of each particular mission, the Civil Engineering Center at Wright-Patterson AFB, Ohio, monitored the overall activities of Project RED HORSE in accordance with AFR 85-32. Hq USAF directed all squadron deployments.[7]

Organization

According to AFR 85-25, RED HORSE Squadrons were self-sufficient to a limited degree and provided the following services:[8]

- Food and medical service, heavy vehicle and equipment maintenance, basic supply functions (AFM 67-1, Chap II, Part 1, Vol I).

- An engineering staff to plan and direct expeditionary airfield work and emergency airfield repairs.

- Site development teams equipped for land surveys and soil tests and experienced in selecting, designing, and preparing expeditionary airfields and supporting facilities.

. A work force with sufficient construction materials and equipment to place expeditionary airfields and supporting facilities into operation and to perform emergency and heavy repair work required for sustained combat unit operations.

. Training, management capability, and financial resources to enable immediate augmentation of the CES (HR) work force with local nationals when security conditions permit.

Figure 2 depicts the organizational chart of a typical RED HORSE Squadron.[9]

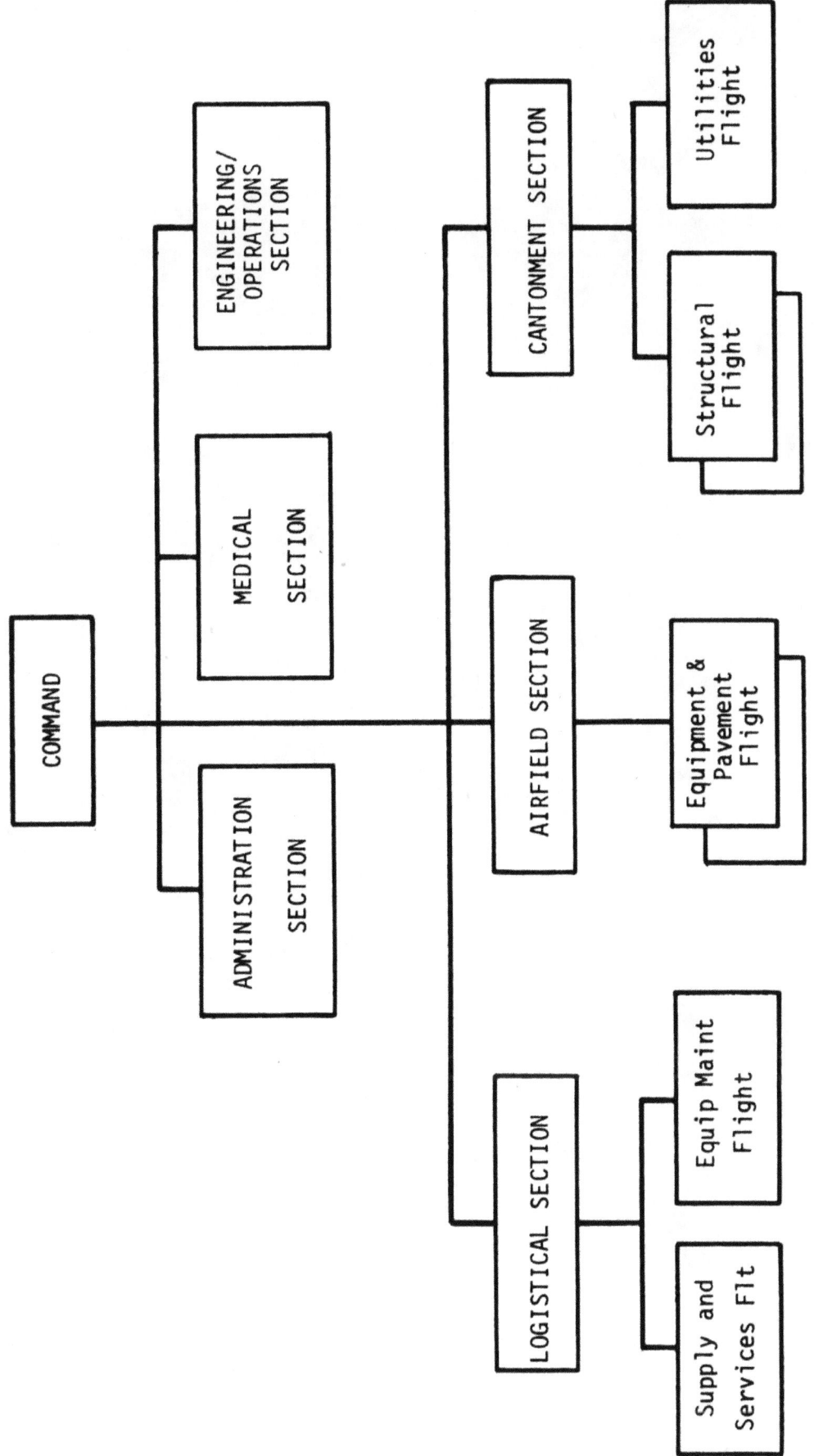

FIGURE 2

CHAPTER II

RED HORSE ORGANIZATIONS IN SOUTHEAST ASIA

Introduction

The following section contains discussions of each major RED HORSE installation in Southeast Asia. (Fig. 3.) Included in each discussion are: a brief history of the unit; major accomplishments; where applicable, timely response to bomb damage or other combat actions; and special discussions on such things as the CONCRETE SKY project, Civic Action, the training of Vietnamese civilians, and other programs unique to a particular base.

Phan Rang - 554th CES (HR)

> *"An advanced party of the 554th CES arrived at Phan Rang on 8 January 1966 and prepared a temporary cantonment area. The entire squadron was in place by 30 January. It had completed construction of its cantonment area by June 1966."* [1]

The arrival of the 554th Squadron at Phan Rang was both similar and unique in relation to the beginnings of other RED HORSE operations in Southeast Asia. (Fig. 4.) It was similar in that most squadrons sent an advance group of men to prepare the site for the arrival of the total squadron. These men prepared for the construction of living space, a work area, and came to grips with whatever problems might hinder the initial deployment. The 554th Squadron was unique in that it was one of the first two squadrons to arrive in Southeast Asia.[2] Other squadrons which followed would learn and benefit by the experience of this early squadron. Eventually, the unit deployed men to Phu Cat, Dalat, Da Nang, and in September 1968, activated a unit at Binh Thuy, which replaced a similar unit previously furnished by the 823d CES.

One of the first tasks assigned to the squadron was the repair of AM-2, mat-surfaced runways which had been damaged by aircraft accidents. To avoid heat exhaustion and interruption of air traffic, the work on this high priority, combat support item was done primarily at night. The AM-2 replacement task had to be continued during the Southwest Monsoon Season, which began in May, because some of the griffilyn membrane installed below the surface mat had ripped. Water, under the pressure of heavy aircraft, eroded the soil beneath the runway and caused depressions in the surface mat. According to some observers, "...soon aircraft were hitting depressions so deep that weapons suspended below their wings dragged on the ground...." The repair project, accomplished almost entirely at night, involved the removal and repair of almost two million square feet of AM-2 mat and the removal of 41,000 cubic yards of damaged base course material.[3]

In the spring of 1967, the 554th became the first Air Force Civil Engineering unit in Vietnam to own and operate a concrete batch plant. (Fig. 5.) This paving capability was still being used extensively at the time this report was written. For example, in March 1969, the 554th concrete batch plant produced 8,310 cubic yards of concrete. Most of this production was used for the construction of large, concrete dispersal hardstands (Fig. 6) and concrete covers for protective shelters. Of the 33 Military Construction Program (MCP) and O&M projects (dollar value 2.4 million) in progress during March 1969, the concrete shelter program was the largest. This project consisted of the construction of 61 concrete covered aircraft protective shelters. With the exception of 16, these were located on new, dispersed hardstands. Shelter

FIGURE 3

554th RED HORSE COMBAT DEFENSE

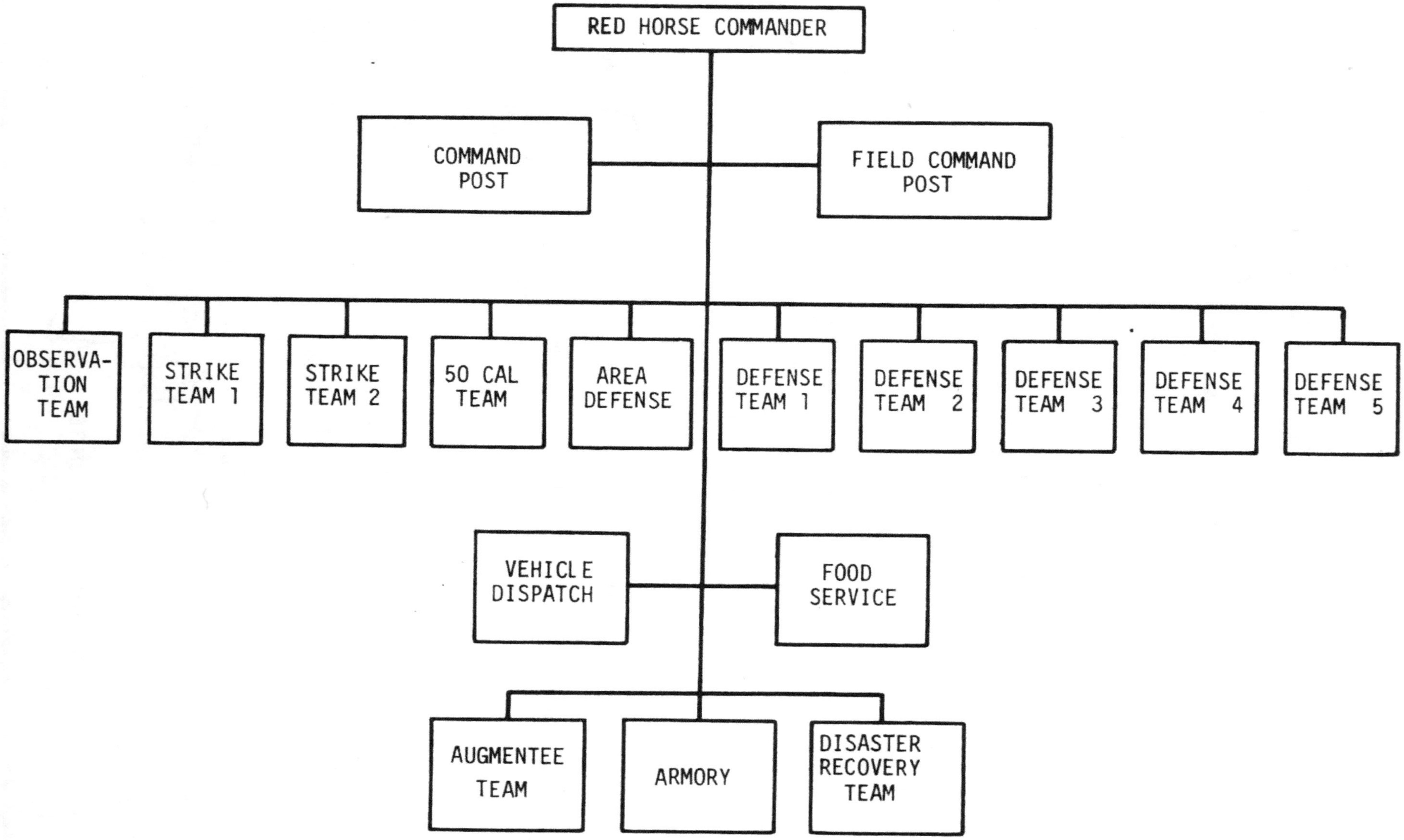

- RED HORSE COMMANDER
  - COMMAND POST
  - FIELD COMMAND POST
  - OBSERVATION TEAM
  - STRIKE TEAM 1
  - STRIKE TEAM 2
  - 50 CAL TEAM
  - AREA DEFENSE
    - DEFENSE TEAM 1
    - DEFENSE TEAM 2
    - DEFENSE TEAM 3
    - DEFENSE TEAM 4
    - DEFENSE TEAM 5
  - VEHICLE DISPATCH
  - FOOD SERVICE
  - AUGMENTEE TEAM
  - ARMORY
    - DISASTER RECOVERY TEAM

FIGURE 4

RED HORSE Concrete Batching Plant
Figure 5

Pouring Concrete Lane
Dispersal Hardstands
Figure 6

construction consisted of a series of steel arches, which were later covered with 15 inches of concrete. These shelters proved to be almost invulnerable even when sustaining direct hits from 120-mm rockets. In addition, they provided a somewhat cooler working space for maintenance crews. [4/]

As pointed out in a 554th CES (HR) briefing: [5/]

> "This /aircraft shelter/ project is a classic example of the capability of this RED HORSE Squadron. It involves surveying, earth moving, placement of forms for paving, a quarry operation and rock crushers to produce aggregate for the concrete and asphalt, electricians to install taxiway lighting, steelworkers to erect the shelters, carpenters to place forms over the shelters or concrete...a crane and bucket operation and a concrete pump. (APP. I, Figs. 7-11.)

During the period of 10 August 1968 through 31 May 1969, the 554th completed 41 construction projects with a troop construction dollar value of 1.2 million. Major projects completed and turned over to the 35th Tactical Fighter Wing included an armament and electronics shop, flight line fire station, a concrete access taxiway, six airmen dormitories (Fig. 12), two 20-man officers quarters, 49,000 square yards of asphaltic concrete for base supply, a runway Visual Approach Slope Indicator (VASI) lighting system, a tactical air navigation (TACAN) facility, a Runway Surveillance Unit (RSU), and BAK-12 barrier installations. (Fig. 13.) A new base theater was also built and more than one mile of reinforced concrete revetment wall was erected to protect the living quarters of the 35th TFW and tenant units. [6/] (App. I.)

The employment of Vietnamese nationals increased from 11 in April 1966 to

685 by September 1967. Initially, most were hired as unskilled laborers or as painters, carpenters, and masons and had to be closely supervised. Eventually, a semiskilled labor pool was built up and included vehicle mechanics, rock crushers, and asphalt plant workers and other skilled personnel.7/ (Fig. 14.) Much emphasis was placed on the employment and training of Vietnamese nationals. At the time this report was written, there were 573 Vietnamese civilians employed in 11 different skills. Of this total, 325 were formally enrolled in the OJT program. Each man received at least six hours of formal classroom instruction in addition to his on-the-job training. Further, electricians, plumbers, power production specialists, repairmen, and heavy equipment operators were sent to school for three months in Saigon and elsewhere.8/

According to Sergeant Munci, NCOIC of Phan Rang's RED HORSE civilian personnel training section, "The objective of the training program was to bring a man to the journeyman level of proficiency within 18 months." To this end, Job Proficiency Guides and standards were used and four interpreters were sent to various sections to aid with the training. In addition, those Vietnamese who could neither read nor write, were encouraged to acquire these skills from evening courses given on base. A pay incentive was given to those individuals who could demonstrate a knowledge of English. These skills, of course, aided measurably in successfully accomplishing the mission of the squadron; yet, they did more than that. They provided the country itself with a growing pool of skilled laborers who one day might work in Vietnamese industries. Many Vietnamese nationals had already advanced to supervisory

RED HORSE Defense Line
Figure 7

Forming Aircraft Shelters for Concrete Cover
Figure 8

Covering Aircraft Shelters with Crane/Bucket Method
Figure 9

Covering Aircraft Shelters with Squeeze Crete Pump
Figure 10

Aircraft Shelters with Concrete Cover
Figure 11

Construct Airmen's Dorms
Figure 12

Construct Aircraft Barrier (BAK-12)
Figure 13

RED HORSE Asphalt Mixing Plant
Figure 14

positions in their particular areas. In this respect, RED HORSE training contributed to the enormous task of Nation Building in Southeast Asia.[9] (Fig. 15.)

At Phan Rang, as elsewhere in RVN, RED HORSE was more than a construction unit. It had a combat capability which was frequently used. From 10 August 1968 to 3 May 1969, the RED HORSE Combat Defense Teams were deployed to the perimeter 14 times during mortar/rocket and infantry sapper attacks. Appendix II summarizes the hostile action which necessitated deployment of the Combat Defense Teams.[10]

555th CES (HR) Cam Ranh Bay

The 555th Civil Engineering Squadron deployed to Cam Ranh Bay in February 1966. The construction combine of Raymond, Morrison, Knudsen-Brown, Root, Jones (RMK-BRJ) had completed the airfield and other temporary facilities. The RED HORSE Squadron concentrated on construction of troop housing, roads, utilities, and other support facilities. (Fig. 16.) Although RED HORSE Squadrons had not originally been trained or equipped for this kind of mission, it was recognized that under certain circumstances this kind of construction mission would be required.[11]

At first, materiel shortages were the primary causes of work slowdowns and stoppages. Delays of this nature were common during the early squadron deployments. Equipments deployed with the squadrons had been taken from various bases in the United States and were in varying operating conditions on their arrival in Southeast Asia. Furthermore, material and equipment were often designed for construction missions inappropriate for the particular SEA base at which they arrived. Procurement of common construction materials requisitioned by the 555th CES often took from six to eight months. "In March 1967, the squadron started receiving consolidated supplies shipments from AFLC through Project PACER OAR." This nickname was given to shipload lots of supplies and equipment shipped to SEA for RED HORSE usage.[12/]

In time, the logistics section was upgraded and staffed to accommodate the additional workload caused by this shift in procedures. By October 1967, the 555th CES had completed more than 100 projects at Cam Ranh Bay...the squadron had constructed more than 100 buildings (totaling over 480,000 square feet), completed more than 650,000 square yards of earthwork, and built 25,000 linear feet of utilities. In June 1967, the squadron detached an 18-man unit to Phu Quoc Island in the Gulf of Thailand to construct radar facilities. In the same month, it deployed a unit to Nha Trang to replace the deployed unit of the 820th CES. By October 1967, the deployed unit had completed six projects and was working on 10 more. The dollar value of the completed projects at all locations exceeded 10 million.[13/]

The 555th, as did other squadrons in SEA, experienced some problems with AM-2 runway matting. AM-2 maintenance was a base function but was assigned to the squadron, because it had the necessary manpower and equipment resources. This maintenance was performed mostly at night to avoid heat exhaustion and interference with air traffic. Early in 1967, the matting showed the effects of shaping by heavy aircraft movement. There were broken panels, subgrade failures, and washouts beneath the surface. Approximately 420,000 square feet of runway and taxiway were repaired in 1967.[14/]

Constructed Railroad Spur
Figure 15

POL Pipeline
Figure 16

Some major projects completed in FY 68 included: an open storage, base fumigation chamber; construction of a radio beacon facility; installation of an aircraft arresting facility (BAK), and construction of temporary revetments. (Fig. 17.) During the first six months of 1969, the 555th completed 29 projects including new C-130 crew quarters, a chapel, a base exchange snack bar, a clothing sales store, and a 4,000-square foot recreation workshop.[15/] (App. III.) The 555th also made important contributions to the local economy by employing and training Vietnamese nationals. At one time, as many as a thousand people were on the payroll, but this number had dropped to 620 at the time this report was written. In training these civilians, a controlled system of evaluation and upgrading was utilized. Job proficiency guides and standards were used, as well as formal classroom training, trade schools, instruction in Vietnamese reading and writing, and a pay incentive for knowledge of English. Eighty people had been upgraded in the year preceding the writing of this report. In addition, many employees who demonstrated a high job proficiency were selected to attend the Base Management Course in preparation for supervisory positions.[16/]

In addition to utilizing local workers, several men of the squadron were active in the Civic Action program during their off duty hours. Cement spillage and other materials unfit for use on base projects were donated and volunteers worked with the people of the area to complete various building projects. In the past year, they contributed to projects such as: CONCERN, Cam Ranh orphanage, My Ca Refugee camp, and the construction of a playground for the Ba Ngnoi school. Emphasis was always placed on helping a particular group help themselves rather than to supply material or do the job for them.[17/]

The squadron also provided medical care to Vietnamese workers. In the first quarter of 1969 alone, there were 994 civilian visits to the dispensary, and a total of 2,104 immunizations (including military and civilian) administered.[18]

819th CES (HR) - Phu Cat

The advance staff of the 819th arrived at Phu Cat on 6 August 1966, and was joined by personnel from the 555th and 554th to complete beddown facilities for the main body of the squadron. It was estimated that the efforts of the advance staff, the planning and positioning of materials at Qui Nhon by the Seventh Air Force, the assistance of the 554th and 555th CES, and the employment of Vietnamese civilians, saved the squadron at least 30 days in getting six base projects ready.[19]

Because civilian contractor personnel began to phase out after the first year, several construction jobs, installation of the overhead distribution (power) system, water and sewer lines, and area paving had to be finished by RED HORSE personnel. Phu Cat became the one base in the Republic of Vietnam where almost all building and construction, and the greatest percentage of earthen and paving construction, were accomplished by RED HORSE.[20]

During the first year of operation, the 819th was logistically supported by a BITTERWINE B1-C package and LOGGY STEED 131-E package. It also received materials requisitioned by Seventh Air Force, as well as buildings procured by the AFLC in the CONUS and Seventh Air Force in Singapore.[21]

As with other squadrons in SEA, the 819th experienced a time lag of up to 10 months in the arrival of CONUS materials. The squadron logistics

Revetments, F-4C
Figure 17

operation performed the function of the Base Accountable Supply Office (BASO). Demonstrating the design potential common to all RED HORSE squadrons, the design section of the 819th developed its own standard designs based on ease of construction and available materials.

The placement of AM-2 matting presented problems to the squadron that were similar to those at other bases. Problems of thermal expansion and contraction, base and subbase failure, and membrane damage appeared again and again. The matting was quickly repaired by several crews, each consisting of 10 airmen and 50 Vietnamese. By working shifts, these crews replaced as much as 75,000 square feet of matting at a time. [22] Additional accomplishments by the end of October 1967 were: [23]

> "The 819th had moved 1,659,000 cubic yards of earth, poured 15,500 cubic yards of concrete and constructed buildings totaling 633,000 square feet. In addition, it had finished over 50,000 linear feet of utility lines, fences and storm drainage facilities, erected over 5,000 linear feet of aircraft revetments and completed over five miles of roads."

During the period between June 1968 and 31 May 1969, there were 23 O&M and 9 MCP projects completed for a total cost of $1,804,200. Some of the major construction operations included a 250,000-gallon potable water storage tank, a security fence (5,000 feet of chainlink, 8 feet high, constructed in 17 days), aircraft ramp replacement, new officers quarters, a VASI lighting system, a BAK-13 arresting barrier system, a communications facility, and an aircraft engine trim pad. [24]

"The 819th CES response to a major operational mission change posing an especially challenging timetable for facility requirements was a clear-cut example of RED HORSE construction capability. The mission change involved a complete weapons system conversion from F-100 to F-4D aircraft. Working closely with the 37th Tactical Fighter Wing and the BCE, a 'prior to arrival' facility requirements list was formulated, priorities were established, designs were initiated for some facilities even prior to completion of programming action, and construction was initiated for some facilities even prior to completion of programming action, and construction was initiated, in some cases, with verbal approval to proceed from the 1 CEG. Designs were based on existing material resources, and the Critical Path Method (CPM) was employed to insure that a tight schedule was maintained. 'Prior to arrival projects' included the following:

FACILITY	REQUIRED BOD
AMS Shop Modification	*15 April 1969*
High Value and Classified Storage Warehouses	*15 April 1969*
OMS Buildings (2)	*15 April 1969*
Squadron Operations Building	*1 April 1969*
Officers Quarters	*15 March 1969*

"Without exception, the construction schedule was met and the new weapons system became operational at Phu Cat according to plan. In addition to the foregoing facilities, a follow-on list of requirements was identified and some were completed prior to the end of the tour. 25/

The Concrete Aircraft Shelter program at Phu Cat began in December 1968 with the erection of the first row of eight shelters. Concrete cover (15 inches thick) placement began in January 1969 and was completed during the first week in March 1969. The second row of shelters was turned over in mid-April, and the third row of eight shelters on 1 June 1969. The target date for completion of 40 shelters was August 1969. 26/

At the squadron's detachment at Pleiku, significant projects included the reinstallation of BAK-12 Aircraft Arresting Barriers, installation of an asphalt batch plant, installation of a sewer line, construction of an Officers Open Mess (10,000 SF), and repairs to rocket damaged facilities (A&E Facility and Officers Quarters). (Fig. 18.)

Construction Summary

    Buildings/Facilities   161,300 SF
    Aircraft Apron   42,000 SY
    Aircraft Revetments   42 EA
    Asphaltic Concrete Produced   12,000 TN
    Asphaltic Concrete Pavement Placed 150,000 LF
    Sewer Line   8,000 LF
    Open Drainage   2,000 LT

Total $ value of construction completed: 1,100,000.00. [27]

At the time this report was written, 534 civilians employed at Phu Cat with 130 at Pleiku, making a total of 664. In addition to the training of these civilians, the squadron also participated in a training program for the ARVN forces. [28]

> *"In cooperation with Lt. Col. Vu Duy De, Commanding Officer of the Regional Force Training Center, Colonel Marcus Horton, 819th Commander, initiated a training program for 20 ARVN soldiers covering nine job skills. Among the subjects taught during this two-month period, were carpentry, masonry, plumbing, electricity, sheet metal work, power production, vehicle maintenance, and construction drafting."* [29]

Thus a concrete example of cooperation between forces of the U.S. and Republic of Vietnam within the RED HORSE domain was realized. Aside from the Nation Building aspects of the normal civilian training program, the 819th reached beyond them to achieve a common goal.

One of the highlights of the training program at Phu Cat was the implementation of a management course developed by the University of Michigan and modified to fit the needs of the Vietnamese within the framework of their given mission and local situation. As in the other squadrons, Phu Cat administered medical attention to its Vietnamese workers. A Vietnamese nurse and two Vietnamese medical technicians worked with the base doctor and supported his Medical Civic Action Program (MEDCAP) activities in the surrounding community.[30/]

Phu Cat's RED HORSE team also supported the base in its perimeter defense. Recently, the squadron redesigned its defense plan and rebuilt and relocated bunkers. The bunkers were redesigned to include elevated pads for proper water drainage, covered sandbagged roofs for rain protection, and better access roads for resupply. Fields of fire were extensively cleared of overgrowth. A cache of small arms and hand grenades was discovered during this clearing process. In addition, a new perimeter fence, consisting of two rows of triple concertina wire with tanglefoot in between as a median row, was constructed. At the time this report was written, the squadron had deployed to the perimeter only once.

820th CES (HR) Tuy Hoa/Da Nang

The 820th CES began training on 1 July 1966. The advance staff deployed to Tuy Hoa AB, RVN, on 17 August and prepared a temporary cantonment area. Tents were erected to provide living quarters and shops, and by 9 October 1966, the entire squadrons was in place.[31/]

Rocket Damage to Red Horse
Compound, January 1969
Figure 18

The first task was to build interim facilities which would permit early deployment of the 31st Fighter Wing. This Wing was operational by mid-December 1966. An NCO dining hall had also been completed and other projects were begun during the same period.[32/]

On the whole, materiel support was better than in some of the other squadrons. The squadron received large quantities of supplies, but critical items such as electrical and plumbing equipment were scarce. With the installation of the computerized, standard Base Level Supply System in March 1967, the BASO became an effective supporting agency. In June 1967, the contractor turned over $2 million worth of construction materials to the 820th CES as a part of his phase-out operations. Thus, the squadron was well supplied by the time Project PACER OAR began to function.[33/]

In early 1967, the squadron deployed units consisting of 26 military and 150 Vietnamese to Nha Trang, plus a 20-man PRIME BEEF team. These units erected 61,000 square feet of wooden buildings and 84,000 square feet of pre-engineered metal buildings there. In July 1967, they were replaced by a unit from the 555th CES. The 820th also deployed a unit to Da Nang (approximately half of the squadron personnel) and sent men to augment the airfield construction forces at Cam Ranh Bay.

By September 1967, the 820th CES had accomplished almost 50 percent of all construction completed at Tuy Hoa. Including the work performed by deployed units, the squadron had completed 90 major projects. It laid 175,000 square yards of AM-2 airfield matting, erected 170 aircraft protective revetments, and

17

constructed 120,000 square feet of prefabricated and pre-engineered metal buildings (exclusive of dormitories). The squadron also erected inflatable shelters. It operated a rock crusher 9.5 miles from the base and hauled aggregate through enemy held territory to the base.[34/]

At Da Nang AB, between 1 June 1967 and 31 May 1968, a total of 73 construction projects were completed at a cost of $3,059,000. At Tuy Hoa Air Base, 55 projects were completed at a cost of $1,491,952.46. The squadron, as a whole, completed more than 128 projects costing $4,550,952.00.[35/]

By far the most important project for the 820th CES was the Concrete Aircraft Shelter or CONCRETE SKY Project, which 7AF had given a number one priority. In all, the 820th completed over a quarter of the total shelters programmed for South Vietnam. In February 1969, the squadron headquarters had moved from Tuy Hoa to Da Nang, thus reversing the status of the two units. Construction of the CONCRETE SKY Project began early in July. Being the first squadron in Vietnam to erect such shelters, some difficulties were experienced and overcome by this unit. For example, no experienced personnel were available to erect the shelters and no instructions were provided by the companies that manufactured them. In addition, many of the metal parts had been damaged during shipment, and the arches had been manufactured larger than the spaces available for them in the existing revetments. Despite these and other problems, the men soon cut the erection time for one metal shelter from one week to one day, and the entire project (98 shelters) was completed 52 days ahead of schedule.[36/]

The 1st CEG was tasked by Hq 7AF (DCE) to place 12 large, vitally needed antenna poles atop Monkey Mountain to support the 620th Tactical Air Control Squadron. The mountain was steep and had only a few roads leading to the top. In addition, the site had been set in an abandoned mine field. After investigating the site at the request of the Officer in Charge of Construction (OIC), a civilian contractor declared the estimated cost would be $150 thousand. Seventh Air Force did not have the money for the project.

An on-site inspection led to the final commitment of RED HORSE to the project:[37/]

> *"The 620th TAC Squadron wanted twelve 90-foot long poles placed in a grid on top of the mountain which was almost solid rock. Maj. Thorpe, Operations Officer for the squadron, began the project by designing the preliminary layout of the site. United States Marines cleared land mines out of the area. The RED HORSE site development team did a detailed survey of the site staking out the locations for the new poles. RED HORSE personnel drilled holes in the rock, blasted with dynamite, dug out the debris, using jack hammers and other tools and created holes four feet wide by 12 feet deep. The poles, three feet in diameter at the base and weighing 5,000 lbs. apiece, were set in the holes. Concrete was poured around the poles using the squeeze Crete pump machine. This was difficult because of the steep, approximately 45-degree angle of the embankment. The squadron was given 70 days to set the poles. The poles were set in 35 days."*

An emergency situation requiring rapid response developed on 27 April 1969 when the Marine Ammunition Dump and bomb storage area at Da Nang blew up.[38/] From an interview with Lt. Col. Roy Lemons, 820th RED HORSE Commander, and a narrative by Maj. Roscoe P. Thorpe emerged the following description of the events of that day and recovery operations.

According to Colonel Lemons, the grass fire near the dump apparently began some time during Sunday morning, on 27 April. By 1100 hours, it had spread to the storage area, and bombs began to explode. Since explosions were not uncommon around the Da Nang area, not many people realized what was going on at first. It was not until about 1130 hours that it became obvious to everyone that the scope of these explosions could be disastrous. As Colonel Lemons said: "I realized then that RED HORSE would have a lot of work ahead of it."[39/]

Major Thorpe described the incident in his narrative:[40/]

> *"Large columns of white and black smoke could now be seen near the bomb area, about four miles from the RED HORSE compound. This was to be the beginning of hundreds of explosions and intense fires which would continue for over twelve hours. The concussions from the blasts would continue to completely destroy many facilities within the immediate area and cause extensive damage two miles away. As things worsened, the base sirens screamed a warning that the men were to take cover in the bunkers. Personnel were evacuated from the south end of the air base and the newly completed port facilities on the west side of the runway."*

According to Colonel Lemons, "By 1400 hours, the base had sustained limited damage." Between 1500 and 1900 hours, the base had sustained major damage. Bombs, rockets, ammunition, antipersonnel explosives, napalm, and a variety of other devices were exploding in a chain reaction, each explosion sending a huge shower of sparks, flame, shrapnel, and partially exploded or unexploded portions of ordnance high into the air and into adjoining areas.[41/]

As related by Colonel Lemons:[42/]

> *"At 1500 hours, Major Winters, the Acting Base Civil Engineer, and I met to discuss the situation and began to establish procedures and responsibilities of what was to be a very extensive recovery effort. It was decided that RED HORSE would pick up the heavy damage and virtually all major repairs. It would concentrate on those areas, including Air Force dormitories and the aerial port facilities, on the west side of the base. The BCE would pick up the Security Police complex and would save the utility systems. We made frequent tours of the base as the explosions were taking place, to assess damage and keep abreast of what was happening and what was ahead for us. The air was filled with black smoke and debris and we had an 'early sundown' at 1700 or 1800 because of this."*

In the evening, at approximately 2000 hours, the RED HORSE staff, including officers and key NCOs were called together by the Commander of the 820th for a "brainstorming" and in-depth preplanning and programming session. The blasts were still occurring, sometimes subsiding only to erupt again with even greater intensity and power. The shock waves from the explosions were clearly visible and of violent intensity in the RED HORSE area, four miles from the source of the explosions. The over pressure caused by the shock waves was at least two pounds per square inch in the Air Force area. [43]

During the conferences in the RED HORSE compound: [44]

> *"A huge map was displayed with the damaged base areas outlined in red. First hand damage reports were studied. Ideas were exchanged. Teams of electricians, plumbers, and carpenters who would survey the damage were organized. Work orders were established. Cost accounting procedures and practices were reviewed. Plans were carefully detailed so that the initial evaluators would be fully prepared to enter the area at first light the following morning. Trucks were ready for dispatch."*

The RED HORSE men retired early that night, while the bombs were still exploding and flames and debris were still towering hundreds of feet in the sky. They gladly shared their quarters and living areas with approximately 500 Marines who had been evacuated from the danger area. Because the squadron area had originally been designed to support only 200 men, the entire group of 900 found living a bit cramped. 45/

As Colonel Lemons remarked: 46/

> "...however, the two groups got on very well. They were in a disaster situation. The RED HORSE men had an important mission to prepare for and there was no time nor inclination for incidents of any kind. A spirit of camaraderie emerged which was very inspiring to watch and the bond that was formed that night has continued between the officers and enlisted men of RED HORSE and those of the Marines."

By that evening (approximately 2315 hours), the explosions and fires had subsided. About six hours later at 0515 hours, the Acting Wing Commander of the 366th Tactical Fighter Wing gave permission for RED HORSE to enter the damaged aerial port area. The RED HORSE Engineering Officer in the field (Major Thorpe) hurried to the scene with teams of NCOs to examine the damage and assess the manpower and materials required for the initial repairs. Lt. Gregory Dumas, a RED HORSE Operations Officer, became the liaison officer between the BCE and the RED HORSE unit. In the words of Maj. Roscoe P. Thorpe: 47/

> "The damage was extensive in many of the 24 buildings. Utilities were off in most areas. Ceilings and walls were crumpled and shoved aside by the force of the blasts. Light fixtures, base wires, and broken glass were everywhere. Windows, doors, and wall panels were splintered and broken. Steel members of pre-engineered

> *buildings were twisted and bent.*
>
> *"As with most buildings in the area, the Airlift Control Element was in shambles. The ALCE officer in charge informed the RED HORSE team that his command post was expecting the first plane at 0730 hours. There was no electricity, no radio, no communications, and the windows were shattered. The following plans were detailed the night before the RED HORSE team went into action. Workmen and materials were called for by radio. They arrived by 0630 hours. Within 45 minutes the building was made safe and electric power was restored to the vital communications equipment. The first plane landed at 0730 hours, right on schedule. The passenger terminal and adjacent operations buildings were quickly repaired and made ready for limited operations."*

This was, of course, vital to the operation of the base. During the holocaust, planes had to be diverted or delayed. As a result, traffic stack-ups were experienced all the way back to the U.S. Planes were stacked up at Tan Son Nhut, the Philippines, Hickam AFB, and even Travis AFB in California. When the first aircraft touched down on schedule the next morning, air traffic over half the world started to flow smoothly again.[48/]

As other areas closer to the source of the explosions were cleared by the Explosive Ordnance Disposal team, RED HORSE crews spread out, making immediate repairs to prevent injury to personnel returning to damaged facilities. Everywhere the story was much the same: shattered windows, downed lighting fixtures, roofs and walls buckled, water leaks, structural members bent and twisted. Some buildings were a total loss, others were partially destroyed, but usable.

Colonel Lemons had established a three-phase operations and repair schedule. The first phase would be to establish by building work order number, priorities according to the critical nature of the function, building, and area. The second phase was the reconstruction of important facilities with partial or "functional restoration" of facilities with lesser priorities. Phase three was the complete repair of all damaged facilities.

With critical repairs now accomplished where possible, it was time for RED HORSE to establish a long-range repair program involving a priority system which was consistent with the mission requirements and material availability. Work was stopped on most scheduled RED HORSE projects while emergency repairs were in progress. More damage was uncovered as teams probed farther into the individual buildings. Working closely with the 366th Tactical Fighter Wing, RED HORSE continued with the repair of 68 buildings and facilities with an estimated restoration cost of $1.5 million.

The RED HORSE control center quickly shifted from routine operations to the nerve center for the damage repair. New charts were prepared and constantly updated to show work progress on each building. The result of each detailed inspection was recorded and required materials not available were requested from other RED HORSE units in Vietnam. Available material, such as structural steel members were redesigned and rebuilt to satisfy a particular requirement. Many buildings were stripped to bare skeletons before reconstruction could begin. Ingenious methods were devised to expedite completion of unusual problems. Cost accounting procedures were set up to accurately reflect the daily cost for all emergency projects. Cumulative costs were recorded for periodic review. In less than one month, much of the damage was a matter of record.[49/]

The Commander of the 820th summed up his feelings this way:

> *"I don't have enough adjectives to describe the work of these G.I.s. Everyone of them leaped at the opportunity to demonstrate what they could do when the chips were down and they did it. Within six days I was able to pull one half the work force from the bomb damage repair in order to continue with other needed projects which had been scheduled before the blast. In view of the extensive nature of the damage, this is incredible. Crews are still working in the dump itself with the EOD people, carefully clearing debris and the area is full of unexploded and dangerous ordnance. Everyone of these men working in the area is a volunteer. We had a man wounded by an exploding antipersonnel weapon as he was working and two men asked to be relieved of their duty in the field. Six volunteers came forward to take the place of the three who had left. The kind of morale and esprit de corps these men displayed during and after this day is beyond description."*

Thus, on 27 April 1969 at Da Nang Air Base, RED HORSE fulfilled its greatest mission as a Rapid Response Heavy Bomb Damage unit in SEA.

Training of all personnel, including Vietnamese civilian workers, received continuous emphasis at Da Nang. Very stringent controls and an evaluation system, including standards, guides, and formal evaluations, designed to bring a man to the journeyman level within an 18-month period, were included in this program. Interpreters were used to implement the program and supervisory instruction was also available.

The Civic Action program at Da Nang was hampered because the surrounding villages were not secure. At Tuy Hoa, however, it continued to flourish. English classes were conducted by several members of the detachment, sanitary conditions were improved at local orphanages, and RED HORSE personnel supervised and worked with the local population after duty hours to help rebuild homes which had been damaged by fire, weather, or war. 50/

The 820th Civil Engineering Squadron's Medical Officer, Captain Stewart G. Selkin, voluntarily offered his services as Chief of Medical Civic Actions at Tuy Hoa Air Base, in addition to his normal duties as a doctor. In this position, he planned, coordinated, and supervised the Medical Civic Action program at Tuy Hoa Air Base. The program had three objectives: (1) to provide the inhabitants of local villages and hamlets diagnostic and therapeutic care; (2) to train Vietnamese medical personnel to be competent in taking care of diagnosis and treatment of disease; (3) to establish programs of public health and preventive medicine. Doctor Selkin personally provided care to approximately 1,200 children at Loc Them and Mang Lang orphanages in the city of Tuy Hoa. On Saturdays, he and his assistants visited Dong Tac to provide medical care for 3,500 refugees. He performed medical Civic Action duties on five of the seven days of the week in addition to his normal medical duties at the Tuy Hoa Air Base Dispensary. Doctor Selkin risked his life for the program, and came under enemy fire on one occasion. He repeatedly visited remote Vietnamese villages and Green Beret camps and Montagnard villages to provide medical care for the Vietnamese people.

823d CES (HR) Bien Hoa

The advance party of the 823d CES arrived at Bien Hoa on 23 October 1966 to erect housing for the main body of the squadron which deployed a week later. 51/ Beginning in November 1966, the 823d deployed self-sustained and equipped units to complete approved projects. By January 1967, deployed units were in place at Tan Son Nhut, Vung Tau, Da Nang, and Pleiku.[52/]

There was a special unit at Bien Hoa which was responsible for site preparation, foundation, and placement and revetment erection. In March and June, personnel were deployed to Da Nang to assist in the repair of damage resulting from mortar and rocket attacks. In May 1967, personnel performed similar work at Bien Hoa. By October 1967, the special forces unit had completed 134,000 square yards of earth and paving projects, erected 6,800 linear feet of revetment, and installed 5,400 linear feet of fence.[53/]

By October 1967, the Deployed Unit #1, at Bien Hoa had completed or was constructing buildings totaling 190,000 square feet. Because much of its unit was deployed at Vung Tau, the Deployed Unit #2 at Tan Son Nhut, was augmented by 48 PRIME BEEF personnel and 100 Vietnamese. The Vietnamese worked in teams consisting of one foreman, two leaders, and 20 workers. Each team had at least one Air Force supervisor and one assistant. In early April 1967, the unit consisted of 18 RED HORSE personnel, 57 PRIME BEEF personnel, and 150 Vietnamese.[54/]

By the end of October, Deployed Unit #2 had completed 20,000 square yards of earth and paving projects, 1,300 feet of revetment and 93,000 square feet of building construction and modification, including much of the 7AF complex. In addition, it had sent a team of RED HORSE and PRIME BEEF personnel to erect 2,000 linear feet of revetment at Binh Thuy.[55/]

On 26 November 1966, the 13-man advance staff of Deployed Unit #2A arrived at Vung Tau AB. By the end of October 1967, it had completed 11 projects,

27

including more than 40,000 square feet of building construction.[56]

A 14-man advance staff of Deployed Unit #3 arrived at Da Nang on 27 December 1966. After completing its preliminary work, a full complement of men arrived to begin construction and repair tasks. By the end of September 1967, Deployed Unit #3 had completed 17 projects, including 85,000 square feet of building construction and modification, as well as 4,300 linear feet of revetment. In February, March and July 1967, Da Nang AB was damaged by rocket and mortar attacks. RED HORSE personnel not only responded to these urgent repair needs, but volunteered for other duties, such as driving ambulances to remove wounded Marines. This unit was deactivated on 10 October 1967.[57]

On 26 December 1966, 28 personnel of Deployed Unit #4 departed Bien Hoa for Pleiku AB. Within 10 months of its deployment, the unit had completed 16 projects, including 74,000 linear feet of revetment.[58]

During the TET Offensive and the following period (January 1968 - 31 May 1968), the squadron at Bien Hoa responded to the emergency situation by undertaking some 21 major projects. These included repair of crews quarters, the TASG building, three barracks, the NCO club and other miscellaneous repair jobs. Also, approximately 1,150 man-hours were expended repairing and patching shrapnel-damaged F-100 revetments. The AM-2 ramp for the Hot Cargo area had to be repaired. Overall, more than 300 man-hours were spent replacing the sub-base, the base, and AM-2 matting.[59]

The RED HORSE line construction crew devoted a maximum effort to restore power to the east cantonment area. By 1900 hours of the day following the initial attack, power was restored to all mission essential facilities. Overall, approximately 1,200 linear feet of overhead transmission lines and 1,300 feet of secondary circuitry were installed to restore complete electrical power to this area. 60/

The heavy equipment section of the squadron undertook the following projects during this period: 61/

- Preparation of graves and burial of enemy killed during penetration of the base on 31 January.

- The digging of trenches for defense of the U.S. Army's 145th Aviation Battalion.

- Removal of an earth berm on the ditch along the east perimeter road which the enemy could have used to advantage during a ground attack.

- Heavy equipment was also supplied to clean up the III Corps and VNAF supply area following rocket attack damage.

- 823d Civil Engineering Squadron water trucks supported 3d CES and III Corps in fire fighting operations following the numerous rocket attacks on Bien Hoa AB.

In addition to the normal operations, the squadron also supplied augmentees to the Security Police on an as needed basis throughout the 31 January to 31 May period. From 31 January to 5 February 1968, these augmentees were the first to report and were assigned to key positions which bore the brunt of the attack. For their participation, 17 augmentees were recommended for the Air Force Commendation Medal. 62/

There were 48 projects completed from August 1968 to June 1969, with an in-place value of $3,900,000. The total scope of these projects was: 63/

    Vertical   725,000 SF
    Horizontal   667,000 SY
    Air Conditioning   580 Tons

Base Security absorbed 2,038 man-hours.

1st Civil Engineering Group

At first, the RED HORSE project in South Vietnam was monitored through the 2d Air Division. Such monitoring was carried out to insure that limits of the project were carefully observed and that efforts of RED HORSE Squadrons would not be diverted or directed toward unessential projects. Thus, Hq USAF directed the Commander, 2d Air Division, to control the project. This control was accomplished through a Deputy Commander for RED HORSE within the 2d Air Division's Directorate of Civil Engineering. [64]

Yet, because of the decision to deploy a total of five squadrons to SEA, the Directorate requested permission to form a 1st Civil Engineering Group in June 1966. Manning began in October and the group began to function in November. At first, Hq USAF did not approve the Group and it was reintegrated into the Directorate. A second request was approved, however, and the 1st CEG began operations on 15 May 1967. [65]

The Group was organized with a command section (Commander and Deputy) and three divisions. The Engineering Division standardized certain facilities designed by RED HORSE, reviewed squadron projects, established a master file of all designs for construction in the Vietnam theater, issued construction directives for projects assigned to RED HORSE and maintained liaison with the Air Force Regional Civil Engineer for Vietnam (AFRCE-RVN). This was done to insure proper design and siting of RED HORSE projects, and to formulate specifications and project requirements by reviewing and evaluating contractor proposals for various projects requiring procurement of pre-engineered structures.[66/]

The Operations/Plans Division analyzed squadron capabilities, monitored construction schedules and priorities, and coordinated operations with Seventh Air Force. It also maintained liaison with the Military Assistance Command, Directorate of Construction (MAC-DC) and OICC agencies in-country.

The Materiel Division kept a consolidated list of squadron equipment and resources, expedited urgent supply requests, identified sources of supply and procurement, controlled and directed redistribution of equipment assets in light of the overall RED HORSE mission, and assisted in critical equipment parts and repair actions.

One of the most important things to evolve from the group was the standardization of the design for basic types of wooden structures used in Vietnam. This Standard Modular Design Package, completed in conjunction with AFRCE-RVN, utilized criteria based on field experience and permitted design in four standard widths (20, 24, 32, and 40 feet) with the length varying with the scope of the project. The package also contained a list of federal stock numbers for materials used in such structures. This reduced the design work of the squadrons. Since materials were standardized, squadrons could order well in advance, easing some of the strain on the logistics system. Maintenance was easier since designs were standardized. Finally, standardization permitted efficient use of the unskilled Vietnamese labor.[67/]

556th CES (HR) U-Tapao

In July 1966, the 556th RED HORSE Squadron deployed to U-Tapao AB, Thailand, and by October had detached units at Udorn, Takhli, Nakhon Phanom, Ubon, and Korat. These units encountered problems similar to those of the other RED HORSE Squadrons. For example, there were delays in designing and assigning projects, accumulating materials, and obtaining sufficient or appropriate equipment. Some delay was experienced when the using agencies did not recognize a requirement until close to the "need date." There were also problems having to do with the quality of some local products, equipment and spare parts, and language difficulties. 68/

The number of authorized Thai nationals grew from 764 when the squadron was originally deployed, to 3,133 spaces by mid-1967. Most Thais seemed to have higher skill levels than their Vietnamese counterparts and there were fewer security problems. Eventually, many well-trained and experienced Thais were hired to supervise their countrymen. 69/

According to the "RED HORSE in SEA" interim report: 70/

> *"With the exception of Nakhon Phanom, the 556th was engaged primarily in building construction. Within the first nine months, the squadron had completed 15 million dollars worth of construction. This included hauling and compacting 2.1 million cubic yards of fill, building dormitories for 6,200 men and dining halls serving 1,800 men. The total construction program through October 1967 included 326 assigned projects. Completed construction was valued at 25 million."*

In addition to the operating locations mentioned, there was a design and funding function located at Don Muang. By and large, during the period of this report, Nakhon Phanom absorbed the largest effort. Quantity of work at the remaining sites was in the following order: U-Tapao, Udorn, Korat, Takhli, and Ubon. A wide variety of projects were accomplished during the period of May 1968 to April 1969. Projects included BAK-13 arresting barriers (installed and operational within 10 days), 10,800 square feet of hangars (erected and operational within 45 days), a 2,000-foot taxiway with culverts and drainage (constructed and paved in 45 days), and five B-52 revetments (erected and filled within 10 days). (App. V.)

A major repair project was accomplished at Nakhon Phanom (NKP), as heavy rains and other environmental factors necessitated repair or replacement of large sections of the AM-2 mat-surfaced runway. During the period of May 1968 to April 1969, 68 inches of rain fell at NKP, most of which came during the Monsoon Season (June-September). In spite of a three percent crown in the runway, which allowed water drainage into various outflows and culverts, the water penetrating the mat to the sub-surface was too great to drain properly. Water trapped within the laterite was pumped under tremendous hydraulic pressure, creating voids as deep as six inches beneath the mat. At one point, 600 feet of mat shifted approximately 28 feet from the center line alignment. The NKP runway was closed for approximately 40 days while 3,500 feet of the AM-2 mat were removed and the laterite base was repaired. Repairs consisted primarily of recontouring the lateral cross-section of the runway to 1.5 percent slope, improving the runway profile by placing 24 inches of laterite fill in the low area over the culverts, repairing all bad subgrade areas, and placing 12 inches of compacted rock subbase in the remaining 7,000 feet of the runway. Designs for these repairs and others undertaken for the same project were accomplished at Don Muang with technical assistance and review by AFRCE-THAI. Design of a taxiway was accomplished in the same manner. Construction of the taxiway and runway projects was accomplished by the RED HORSE detachment and the squadron at NKP. Personnel were reduced at all other RED HORSE operating locations within Thirteenth Air Force to support this mission. 71/

As far as design capability at Don Muang was concerned, during the period of May 1968 to April 1969, a total of 103 projects, with a value of $6,805,000 were designed. This office worked toward standardization of design, but experienced some resistance and delays at certain bases. Frame construction, for example, was continually weighed against more permanent concrete block design at some bases, regardless of funds, or available materials. Construction standards which were acceptable at some bases had been declared unacceptable at other bases.

It has been recognized that no uniform design standards, which are firm and definitive, can be established to cover all facilities, contingencies, and personnel interests. Yet, the system, at the time of this report, seemed to place the design agency in an untenable and unpopular position, if any efficiency or construction austerity were to be effected. 72/

Other design problems had cropped up in the area of BCE-RED HORSE relationships. As stated in the Commander's End of Tour Report for April 1969: 73/

> *"BCE design criteria and definition of projects designed for RED HORSE construction has improved, but is still marginal. The concept that once RED HORSE design and construction of a project are approved, BCE responsibility ends is often apparent. With limited manpower available and serving six widely scattered Thailand bases by undertaking design of over 100 projects, RED HORSE engineering personnel cannot visit each base to secure design criteria for every project. Preparation of adequate design brochures and/or criteria must therefore come from the BCE concerned.*
>
> *"Finally, some problems were encountered with OICC controlled designs.*
>
> *"Engineering designs prepared by OICC contracted engineers varied from excellent to poor and, in some case, resulted in serious construction delays. Two such projects of poor design were the water treatment and distribution systems at Nakhon Phanom and the Demineralized Water Treatment Plant at U-Tapao."*

In the area of logistics, the same fundamental problems which plagued other squadrons had not been completely resolved. The problems included lack of procurement, supply reaction to programming requirements, and programming requirements which had been generated too late for normal supply procedures/processes to respond effectively.[74/] As of July 1969, the timing between project programming with required BOD dates and availability of materials was still poor. Pipeline time for most items averaged from 180-300 days to arrive from the U.S.[75/]

Another supply problem existed because adequate and realistic stock levels of construction materials had not been maintained by base supply agencies. Therefore, much time was expended in "supply chasing."

The following summary of logistics problems and proposed solutions was taken from the Commander's End of Tour Report for May 1968 - April 1969:[76/]

> *"An overall evaluation and recommendation for correction of supply problems is beyond the scope of this report, however, it is sufficient to say that our supply methods of supporting construction and major repairs are not efficient, are not timely, and are not responsive to either operational requirements or to the ability of RED HORSE to place construction. Some real effort should be made for theater of operations construction to standardize and limit engineering designs to a reasonable spectrum of construction supplies by properly identifying, cataloging, and quantitizing these requirements so that design of facilities can be built around a specified catalog of supplies. Realizing that compromises will have to be made in engineering design and economy, exact wire sizes, exact air-conditioner capacities, etc., may have to be abridged and the next size utilized, nevertheless, some order could be brought into the construction supply procedures where now chaos exists in our limitless sizes and types of materials."*

The number of local nationals employed by the squadron over the years has varied with the workload. Beginning in FY 69, 2,573 total spaces were authorized with strengths at a particular location varying with job requirements. It was generally found that the skill levels of Thai nationals were quite high and improved even more while employed with RED HORSE. However, it had also been noted that: "No reliable and capable job superintendents were ever identified." During the past year, approximately 400 men had been upgraded. [77]

CHAPTER III
TRAINING OF MILITARY PERSONNEL

The evolution of the training of RED HORSE military personnel is a particularly interesting aspect of the overall program. The object at first was to be highly selective when manning squadrons and then to mold each group into a cohesive unit. From the inception of the program, unity and esprit de corps were drilled into each man. This kind of indoctrination was necessary because of the diverse skills and backgrounds of the different squadron elements. Because the unit was meant to deploy together and be self-sustaining, if any section or element of the squadron broke away from the group, the entire mission might have been thrown into jeopardy. Because of the strenuous nature of the work, for example, food service and medical attention became increasingly important. To keep a unit operational, especially when deploying to different areas in Vietnam, these services accompanied the squadrons and detachments especially when the units were going into a completely new area. Thus, perhaps the most important aspect of the early training, was not necessarily proficiency training, but rather the ability to react as a cohesive unit.

TAC was directed to train the RED HORSE Squadron in 1965 and Cannon AFB, New Mexico, was chosen as an appropriate site. [1/] The 554th and 555th trained concurrently, but as separate units. They practiced working under field conditions and an attempt was made to improve, or even acquire for the first time, certain technical skills. The training program lasted nine weeks and was only partially successful. Training equipment was often unlike that used in SEA; there was only a handful of special training projects at Cannon, and the units experienced a shortage of certain skills required in SEA.

When the facilities at Cannon became overtaxed, the training site was shifted to Forbes AFB, Kansas. The first generation of the 556th, 819th, and 823d RED HORSE Squadrons took place at this facility. The same objectives were stressed--unit integrity, exercise in field conditions, improvement of skills--yet this unit had the benefit of lessons learned at Cannon. They were given more experience under field conditions and special TDY trips for training in such things as rock crushing, vehicle maintenance, and water purification. Some problems persisted, but on the whole, those trained at Forbes were better prepared than those at Cannon. [2]

According to a historical interim report on "RED HORSE in SEA,"[3] problem areas hindered RED HORSE training despite the improvements made at Forbes AFB. To overcome these deficiencies, the 560th CES was activated at Eglin AFB, Fla., and began operations in February 1967.

Under the auspices of the Air Training Command (ATC), approximately 2,400 men per year were trained to keep the squadrons up to strength. The training was still directed toward developing a high level of morale, discipline, esprit de corps, and incorporated various phases of equipment and construction techniques, expeditionary repair and construction, as well as survival, use of weapons, personal hygiene, and other special requirements for survival in a particular kind of environment.[4]

CHAPTER IV

PROBLEMS AND SECOND THOUGHTS

This chapter contains two interviews taken verbatim, one from a RED HORSE user, and the other from the Commander of the 1st Civil Engineering Group. These provocative interviews summarize various problems encountered with RED HORSE efforts in SEA and offer speculation about the future of RED HORSE as an Air Force asset.

Interview with Col. Frank L. Gailer, on 9 July 1969:[1]

> Col Gailer, Wing Commander of Phan Rang Air Base, expressed mixed emotions about the RED HORSE mission at Phan Rang. "The quality of work is not good," he said, "it is outstanding." "As far as morale, esprit de corps, and the ability to respond rapidly to a combat support mission, RED HORSE must be rated with the best units in the Air Force." However, he added that in order to be effective, each RED HORSE unit must recognize that their primary mission is to support the base to which they are assigned. It should be made clear, early, that for construction and security purposes they are responsible for meeting the priorities and program dates established by and with the Wing Commander, as approved by higher headquarters. In the past, some construction problems have arisen because the RED HORSE, operating as a sub-unit of a CE Group, disregarded locally established priorities. Many units on base compete for RED HORSE support; but RED HORSE construction effort should never be provided unilaterally by them just because they "want to help out." Such competition would be virtually eliminated if the USAF insisted that the RED HORSE confine its construction effort to the locally established/approved program. Anything less than this results in a less than total effort being applied.
>
> Furthermore, now that RED HORSE has become established, it should integrate more completely with the other units on base. The esprit de corps which was so vital to RED HORSE in the beginning and which motivates each man to do

the outstanding job that he does, must not be allowed
to segregate this unit from the rest of the base. Today, the tendency is to become an entity unto themselves.
They live, eat, and play entirely in their own area.
This kind of separatism, Col Gailer commented, detracts
from the overall mission of the base and has at times
caused an unhealthy rivalry among different units.

Finally, he said that the most important issue now facing
RED HORSE as an Air Force organization is not what they
have done or are doing now, but rather how to carry back
the experience they have gained and apply it gainfully to
a peacetime situation. Unless the Air Force is willing to
give up these assets and lose this capability and experience, it must search for and find a practical peacetime
mission which will reach beyond the Vietnam war. This
mission, he said, would form a bridge between this war and
future contingencies.

Unfortunately, this mission must work around the enormous
problems of interservice competition (Army Corps of Engineers, Navy Construction Battalions) and competition with
private industry. Still in all, none of these were really
able to do the job in SEA. The RED HORSE has amply demonstrated that it is needed and can do an outstanding job on
very short notice. Col Gailer commented that because of
its varied capabilities--engineering, construction, medical,
security force, etc.--it is perhaps the most welcome of any
organization on a AF Base. It is all the more important,
therefore, that a detailed and acceptable peacetime mission
be developed so that this capability is not lost.

In an interview, Col. Joseph M. Kristoff, Commander, 1st CEG, on 22 July
1969 related:[2/]

*"I think we would be further ahead if, when a new Wing
or Base Commander enters the theater, that within the
briefings which are given relative to his particular
base, that he be given a good briefing as to the role
and mission of RED HORSE. There has been, from the
very inception of the program, a misconception on the
part of the Base and Wing Commanders that simply because
a RED HORSE unit was stationed on the base, that this
capability represented assets which he personally controlled.*

"At no time during the conceptual stages of developing RED HORSE units was it ever intended that the RED HORSE personnel would be integrated with the Base Command structure. The program was developed so that RED HORSE would have a complete mobile capability being able to respond to any emergency whether it be bomb damage, emergency repair work, of any kind or new construction.

"We can all understand and appreciate that the Wing Commander has many projects which he would like to accomplish and looks upon RED HORSE assets as assets which he would like to apply against these projects based upon his own assessment of priorities. However, these priorities do not necessarily complement the planning actions which are being established at the various levels of the staff within the Numbered Air Force.

"There is a general feeling among Wing Commanders that RED HORSE establishes priorities of work. This has never been a function of RED HORSE. The determination of project requirements, the assignment of priorities is purely a Wing-Base function. The Base Commander is the one who chairs the Facilities Utilization Board (FUB). He and his staff come up with their requirements and the staff in turn determines the priorities in which these projects should be satisfied. Once this list is completed, it is forwarded to Seventh Air Force, in the case of Vietnam, where its requirements are reviewed against the total program which, in many instances, is known better at higher headquarters than at the base. There is always planning going on--moving units, moving men, moving equipment, etc--and many times, the staff is planning two to six months ahead. Therefore, the list which the Commander submits will not necessarily complement this virtue of the staff agencies knowing what the priorities are and programming accordingly. Now the Base Commander may not be happy with it or may not agree with it, but the priority list which is finally approved by Seventh Air Force is passed to the 1st CEG and we in turn pass it out to the squadrons for accomplishment.

"Further, a base cannot always accomplish projects in accordance with this list of approved priorities. Perhaps a given base will have 40 projects and the first is a requirement for a pre-engineered building which we do not have on hand. If so, we must procure one and this takes time. In fact, you might go down as low as number 10 on *the list before you find a project that you can begin immediately. The Base Commander's question, of course, is 'Why aren't you working on priorities 1-9?' Thus you must explain the nature of your problems and the implications. If the Wing Commander is well informed of these various problems and the status of these projects, then it is a little easier for him to accept the delay. If, however, a squadron or detachment appears to be arbitrarily jumping around on the priorities without keeping the Commander informed, he can become very unhappy, and I say justly so.*

"When we (RED HORSE) initially deployed to Vietnam, we had a very definite list of priorities. At Phan Rang and Cam Ranh Bay, for example, we helped the contractor finish the airfield, we worked with the contractor to put down the AM-2 runway, taxiway, and then proceeded from this to behind-the-line facilities in direct support of the flying mission.

"If initially the Wing or Base Commander has this capability available to him there is every reason to believe, as we review the projects going on today for example, that the more critical items would never be accomplished or else would be accomplished long after they were needed. There is a point in time when BXs, recreation facilities, Officers Clubs--facilities which we call 'nice to have' items are needed and can be built. The question is: 'Do we construct this type of facility in lieu of the kinds of facilities needed to keep the aircraft flying?'

"When I go back and view the period of time when I was here (SEA) three years ago, I find that there were many projects considered urgent at the time which today have not been accomplished. For example, hard core power requirements and some road requirements. At many of the bases, we are still attempting to complete utility projects such as providing a good source of water and sewage treatment.

"One of our major, and one of our oldest, operating bases today does not have a complete water system to service the base. One of our major installations is still rationing water. Yet we had in RED HORSE, if the priorities had been properly assigned, the capability to accomplish these things two or three years ago.

"Another of the things I'm concerned about is a recent change in the supply system. We have developed a very good system called BOM (Bill of Materials) which has been programmed on a quarterly basis to support the types of programs which have been assigned to RED HORSE. September will be the last month in which RED HORSE will receive a BOM. Now we will be integrated with a Base Supply System which programs on the basis of past requirements. This type of system may not be responsive to the kind of work we do.

"A further problem is in the area of logistics and vehicle maintenance. The system for identifying standards for use and condition of vehicles developed by AFLC is appropriate to a peacetime situation but may not be realistic in terms of the mission in SEA. The age of a vehicle, for example, cannot really be judged on a purely chronological basis. Utilizing a two shift, 20-hour work day in some cases, we use vehicles two or three times as much as in a given time in the states. Further, the terrain is rougher and other factors of the environment in SEA make the number of miles on a vehicle also a questionable indicator or standard of expected use. Where the AFLC may provide a standard of five or six years for the utilization of a vehicle, we may get only a third of that time.

"In answer to critics who say that RED HORSE units should be integrated with other base units, I submit that this kind of integration is contrary to the entire philosophy of RED HORSE as a fully mobile and self-sufficient unit. To accomplish its given mission, it must retain a degree of autonomy. When these units deploy to the field, leaving the home base, let's say, they do work together, live together, play together, if you will, and a comradeship develops which is quite similar to some of our tactical fighting units. I certainly would not do anything to detract from this kind of unit integrity.

"Looking ahead toward a peacetime mission, I think it would be disastrous if we did not have planning going on at the present time in the various parts of the world where they have been approved for stationing purposes to have a program ready for them on moments notice.

"Such a program might be very similar to those we have in the United States which we call BEMAR or MERP programs. MERP being Minimum Essential Repair Program and BEMAR being Base Essential Repair Program. It might be that the application of RED HORSE assets toward reducing the amount of projects and funds needs in this area (overseas) would be a good program.

"At the same time, I know, there will be an assessment made of the entire philosophy of RED HORSE operations. I am becoming more convinced that there should be a kind of three-phase program. Initially, there has to be a hard core, approved program which may necessitate moving RED HORSE units around, maintaining a home base. Then after perhaps a year, there might be a military/contractor mix--this being the second phase. At that period, the squadrons and bases would become a little more stable. Finally, and this is something which has not been resolved today in Vietnam, the question arises: 'When do we begin to phase RED HORSE units out?' If there is not a plan, coupled with approved construction programs which designates that at some point in time RED HORSE units can be withdrawn, then we begin to construct second generation facilities. A decision must be made at a point other than at Base level, as to whether these facilities are really required or if RED HORSE units might be better utilized at another location. I think perhaps we are rapidly approaching this point in Vietnam."

CHAPTER V

CONCLUSION

How does one measure the success or failure of a program which has so many different facets and variables? What criteria should be used? First, the question of whether the project accomplished its mission as outlined in AFR 85-25 must be answered. From the preceding chapters, it can be seen from the scope of the building and repair programs and the comments of various people, that RED HORSE definitely exceeded all expectations, and it has certainly accomplished the mission for which it was designed. In recognition of this, all RED HORSE units in Vietnam and Thailand have been awarded the Air Force Outstanding Unit Award. Certain squadrons have received two consecutive awards and one or two are being considered for a third. The entire Vietnam group is being recommended as a single unit.

Yet, once this fact of outstanding accomplishment is established, then questions are raised as to the cost of the program in terms of time, men, and money. Because of its mobility and flexibility, the squadrons have been able to expand or diminish the scope of their deployment, achieving an efficient use of authorized strength of both military and civilian personnel. However, a completely objective and controlled comparison with civilian contractor service or peacetime operations would be very difficult to make at this time. The reason is because, as was stated earlier in this report, from the inception of the program it was recognized that RED HORSE would be used in a combat theater, and there would be many unknown conditions which would be hard to control and, also, it was a totally new program. Therefore, it was never intended that the peacetime cost accounting or work control procedures would be applied against the program. Instead, the Major Command and the Numbered Air Force, through the use of inputs from the squadrons and 1st CEG, developed standards and guidelines as the program evolved. For this reason, certain data are incomplete and even complete sets of data have many variables--such as particular materials, specifications, and construction techniques and equipment--which have not yet been identified or controlled. To attempt a comparison on the basis of data such as these would be meaningless, if not impossible.

As this report was being written, a new stage in the development of the RED HORSE program had begun. A consolidation was beginning to take place. New procedures and methods which were, in fact, similar to or the same as those used on a peacetime military base, were being developed and instituted. In 1969, for example, the first IG inspection of RED HORSE units in South Vietnam was conducted. The 1st CEG had recently instituted cost accounting procedures among the squadrons which should yield usable data for definitive cost studies and comparisons. Within a year, it is expected that enough of these data will have been collected, so that a complete and accurate picture can be drawn.

There are two other aspects which are usually overlooked but which are second only to RED HORSE's role in direct support of the flying mission. The first of these is the enormous role RED HORSE played in the training of Vietnamese civilians. Over the years, RED HORSE provided valuable training in a dozen different skills to thousands of men and women in Vietnam alone. This training, as has been pointed out, has developed into a highly controlled and effective system, incorporating OJT, classroom training, and special schools.

The intrinsic value of this program in Nation Building and cultural exchange has been invaluable. This sort of activity went beyond the support measured in terms of dollars, material, or even military or political assistance. It gave these people a trade which they could use in their own industries long after RED HORSE has left.

As Gen. John P. McConnell, USAF Chief of Staff, stated in the February 1969 issue of the "Airman" magazine:

> "I had a personal experience recently that may illustrate the potential benefits of small unit activity to the people of a developing nation. Last November, while visiting Tuy Hoa Air Base in Vietnam, I watched a group of Vietnamese women, dressed in their traditional black pajamas, mass-produce bricks under the guidance of a few airmen from the local RED HORSE Civil Engineer Heavy Repair Unit. The airmen could not speak Vietnamese and the women did not understand English. Yet there was no communications gap. Through gestures and demonstrations, the airmen got the women to shuffle and mix the sand and to operate the jerry-built brick presses. By modern standards, the process seemed primitive and inefficient, but somehow the bricks kept piling up at an incredible speed and at a surprisingly low cost.
>
> "The operation could by no means be classified as 'civic action' or 'community relations.' The Vietnamese were working for pay, and the major purpose was the production of badly needed building materials. But I am confident that this productive display of mutual cooperation will result in economic and social benefits well beyond the mere completion of a construction project. I would not be surprised to learn in a few years that these same Vietnamese people, with better equipment and improved methods, had developed a thriving business making building materials. What is more important to us is that the local people will probably retain a lasting image of the American military man as a symbol of construction rather than destruction.

"We are living in an age in which dissidence born in poverty, ignorance, disease, and social deprivation can expand into insurgency and open conflict. Despite the fact that the primary job of the Air Force is to prepare for combat and fight if necessary, we have an obligation to assist in preventing insurgency as well as defeating it."

The second important effect that this program has had is on the overall Civil Engineering program in the Air Force. It proved that the Air Force did, in fact, have the capability to respond to a heavy repair or construction mission. Moreover, the concept of a self-sustained, rapidly deployable unit which could, if needed, construct expeditionary facilities was validated. In doing so, Project RED HORSE solidified and set the example for the entire Air Force Civil Engineering program. During interviews with RED HORSE officers, NCOs, and airmen, it was very apparent that the program had increased their personal capabilities, broadened their experience, and motivated each of them in an extraordinary manner. With few exceptions, when these individuals return to the U.S., they will carry with them experience, knowledge, and motivation which they would not have received in the normal USAF Civil Engineering career field. This is true because for the most part, the peacetime mission in the United States is merely planning, programming, and maintenance. The actual construction effort is minimal and, of course, for good reasons. Problems, such as government competition with private industry, labor unions, and others make such a large construction mission unfeasible. However, knowledge and practical experience in this area complement and support the Air Force programs which are in effect in the United States. For these reasons, most people involved in the program believe that some way must be found to translate the present RED HORSE mission into an acceptable peacetime operation, so that these important aspects of the overall program will not be lost.

Although this report describes and documents several different aspects of the RED HORSE program, it raises a number of questions which still need to be answered. For example, when the cost data which are being collected during this fiscal year (FY 69 - 70) are in, an in-depth study and comparison should be made to evaluate RED HORSE projects with civilian contractor and peacetime projects. Further, if specific time savings are to be documented (such studies in any kind of controlled and definitive form are unavailable now), then certain specific projects such as CONCRETE SKY, should be identified and time studies undertaken while the work is underway. Next, the full implications of the training of foreign nationals should be the subject of a single, in-depth study. The feasibility of translating the RED HORSE program into a peacetime situation needs to be studied.

FOOTNOTES

FOREWORD

1. AFR 85-25, "Project RED HORSE," 27 Nov 67, pg 1. (Hereafter cited: AFR 85-25.)

2. Article, Maj Gen Robert H. Curtin, Director, AFCE, "Air Force Civil Engineering Magazine," Nov 67, pg 6. (Hereafter cited: AFCE Article.)

CHAPTER I

1. AFR 85-25.

2. Ibid.

3. AFCE Article, pg 6.

4. CORONA HARVEST Interim Hist Rprt, "RED HORSE in SEA 1965-1967," undated. (Hereafter cited: "RED HORSE in SEA.")

5. AFR 85-25, pg 2.

6. Ibid, pg 2.

7. Ibid, pg 3.

8. Ibid, pg 2.

9. Ibid.

CHAPTER II

1. "RED HORSE in SEA."

2. Ibid.

3. Briefing, 554th CES (HR), Official Visit of Colonel Bower, Spring 1969. (Hereafter cited: 554th Briefing.)

4. Ibid.

5. Ibid.

6. Ibid.

7. Interview, 1st Lt Derek Willard, Historian, with Sgt Munci, NCOIC, Civilian Personnel Tng, 554th CES (HR), 14 Jul 69.

50

8. *Ibid.*

9. 554th Briefing.

10. *Ibid.*

11. *Ibid.*

12. "RED HORSE in SEA."

13. *Ibid.*

14. *Ibid.*

15. "RED HORSE in SEA."

16. Project List, 554th CES (HR), 1969.

17. Interview, 1st Lt Derek Willard, Historian, with 1st Lt Fitzsimmons, Administrative Officer, 555th CES (HR), 18 Jul 69.

18. Monthly Civic Action Rprt, 55th CES (HR), undated.

19. Hist Rprt, 555th CES (HR), Jan-Mar 69, pg 32.

20. "RED HORSE in SEA."

21. *Ibid.*

22. *Ibid.*

23. *Ibid.*

24. *Ibid.*

25. End of Tour Rprt, 819th CES (HR), Summer 1969, pp 5-18.

26. *Ibid*, pg 9.

27. *Ibid*, pg 12.

28. *Ibid*, pg 18.

29. *Ibid*, pp 24-25.

30. *Ibid*, pg 25.

31. *Ibid.*

32. "RED HORSE in SEA."

33. Ibid.

34. Ibid.

35. Ibid.

36. Briefing, 820th CES (HR), 27 May 69.

37. Award Narrative, 820th CES (HR), Nomination for Outstanding Unit Award, 1969. (Hereafter cited: Award Narrative.)

38. Ibid.

39. Ibid.

40. Interview, 1st Lt Derek Willard, Historian with Lt Col Ray Lemons, Comdr, 820th CES (HR), Da Nang AB, 21 Jul 69. (Hereafter cited: Lt Col Lemons Interview.)

41. Unpublished Article, Maj R. P. Thorpe, "RED HORSE Crisis and Response," Spring 1969. (Hereafter cited: Major Thorpe Article.)

42. Lt Colonel Lemons Interview.

43. Ibid.

44. Major Thorpe Article.

45. Ibid.

46. Lt Colonel Lemons Interview.

47. Major Thorpe Article.

48. Lt Colonel Lemons Interview.

49. Major Thorpe Article.

50. Award Narrative.

51. "RED HORSE in SEA."

52. Ibid.

53. Ibid.

54. Ibid.

55. Ibid.

56. Ibid.

57. Ibid.

58. Ibid.

59. Award Narrative.

60. Ibid.

61. Ibid.

62. Ibid.

63. Ibid.

64. "RED HORSE in SEA."

65. Ibid.

66. Ibid.

67. Ibid.

68. Ibid.

69. Ibid.

70. Ibid.

71. Ibid.

72. Ibid.

73. End of Tour Rprt, Comdr, 556th CES (HR), Apr 69.

74. Ibid.

75. Ibid.

76. End of Tour Rprt, Comdr, 556th CES (HR), May 68 - Apr 69.

77. Ibid.

CHAPTER III

1. "RED HORSE in SEA."
2. <u>Ibid.</u>
3. <u>Ibid.</u>
4. <u>Ibid.</u>

CHAPTER IV

1. "RED HORSE in SEA."
2. <u>Ibid.</u>

APPENDIX I

PROJECTS COMPLETED FROM 1 SEPTEMBER 1968 TO 20 MAY 1969

PROJ NO	TITLE	DESCRIPTION	COMPLETION	TOTAL COST
PHR 79-8	Construct Supply Point	Surface 130,000 SF unimproved storage area (2" asphaltic concrete)	Sep 68	$14,532
PHR 71-8	Construct Weather Radar Tower	Construct Tower 40' high	Oct 68	$ 2,348
PHR 148-8	Construct Apron, Access	Construct 3" asphaltic concrete surface for 9,400 SY of apron access	Oct 68	$12,600
PHR 6-9	Road, AC Treatment	Paving of approximately 5 miles of road	Oct 68	$49,454
PHR 150-8	Air Condition Air Police	Install 10 tons of Air Conditioning	Nov 68	$ 7,128
PHR 65-8	Construct Squadron Operations	Construct a 112'x32' wood frame structure on a 4" slab	Nov 68	$16,021
PHR 148-7	Repair Shoulder Stabilization	Repair of 25,556 SY of D.B.S.T. taxiway shoulders	Nov 68	$16,056
PHR 163-7	Construct Road, Perimeter	Construct 1.2 miles of 12' wide gravel road	Nov 68	$15,592
PHR 180-7	Refuse Disposal Facility	Excavation of six (6) 1000'x40'x15' sanitary land fill pits	Nov 68	$20,911
PHR 145-7	Security Fence	Clear 25' wide area for 9 miles and stretch 2 ea rows of concertina wire	Nov 68	$30,485
PHR 141-7	Construct Road, Perimeter	Construct 3.2 miles of 12' wide gravel road	Nov 68	$34,573

PROJ NO	TITLE	DESCRIPTION	COMPLETION	TOTAL COST
PHR 35-8	Repair Water Mains	Construct 16,000 LF of 12" raw water main & 4 ea river intakes	Nov 68	$92,806
PHR 316-9	Construct Supply Warehouse	Construct 4,000 SF pre-engineered warehouse facility	Nov 68	$ 6,826
PHR 150-8	Construct Revetments, Passive Defense	Construct 700 LF of ARMCO revetments	Dec 68	$ 1,907
PHR 321-8 (MCP)	A&E Shop	Construct 7,000 SF of pre-engineered metal building	Dec 68	$61,137
PHR 329-8	Fire Station	Construct 10,000 SF of pre-engineered metal building	Dec 68	$60,455
PHR 94-8	Construct OQ Men	Construct 30'x120' wood frame structure on a 4" slab	Dec 68	$31,990
PHR 132-8	Construct OQ Men	Construct 30'x120' wood frame structure on a 4" slab	Dec 68	$31,990
PHR 95-8	Construct Apron	Place 8,900 SF of 12" PCC concrete on a prepared base	Dec 68	$30,196
PHR 114-8	Construct Revetments, Passive Defense	Construct 1,060 LF of 8' high ARMCO revetment	Dec 68	$ 4,560
PHR 123-8	Construct Taxiway Access	Place 13,700 SF of 12" PCC concrete on a prepared base	Dec 68	$42,597
PHR 78-8	Construct Foundation, Weighing Scale	Construct Concrete foundation and approaches for 1 set of weighing scales	Dec 68	$ 2,038
PHR 21-9	Construct Airmen's Dorm	Construct 4,416 SF of wood frame structure on a 4" slab	Dec 68	$37,581

PROJ NO	TITLE	DESCRIPTION	COMPLETION	TOTAL COST
PHR 34-9	Construct Airmen's Dorm	Construct 4,416 SF of wood frame structure on a 4" slab	Dec 68	$37,581
PHR 146-8	Construct TACAN	Construct concrete hardstands (2 ea) & access A/C road	Jan 69	$ 8,123
PHR 26-8	Road, Access (RSU)	Construct 2 ea 12' wide asphaltic concrete access road, 2 ea concrete pads 6" thick 30' wide by 38' long and 2 ea concrete pads 6" thick 18' wide by 65' long		$ 7,034
PHR 13-9	Lighting Approach, VASI	Install visual approach slope indicator lighting system on North and South concrete runway	Jan 69	$11,996
PHR 126-8	Construct Open Storage, Base	Place 49,000 SY of 2" asphaltic concrete on rock base	Jan 69	$26,087
PHR 73-8	Construct Sanitary Sewer Main	Paved 2,700 SF of asphaltic concrete road, install water and sewer lines to 34 trailers	Feb 69	$12,645
PHR 142-8	Construct Airmen's Dorm	Construct 4,416 SF wood frame structure on a 4" slab	Feb 69	$29,647
PHR 65-9	Repair Revetments	Construct 5,780 LF of 10" thick by 4' high concrete revetment wall	Feb 69	$37,717
PHR 119-7	Construct Theater	Construct 400 seat Base Theater	Mar 69	$50,538
PHR 22-8	Construct Vehicle Maint Shop	Construct Maint Shop Vehicle 4,000 SF pre-engineered building	Apr 69	$ 8,682
PHR 156-7	Construct Fuel Vehicle Station	Construct Fuel Vehicle Station 10,000 BBL with Fuel Pumps	Apr 69	$11,441

PROJ NO	TITLE	DESCRIPTION	COMPLETION	TOTAL COST
PHR 113-9	Construct Supply Warehouse	Construct 4,000 SF pre-engineered warehouse facility	Apr 69	$ 6,900
PHR 178-7	Road, A/C Surface Treatment	Repair Base course and overlay 245,600 SF of road surface	Apr 69	$191,200
PHR 87-9	Construct A/C Barrier (BAK-12)	Install 1 ea BAK-12 Aircraft Arresting Barrier	May 69	$ 15,800
			Total	$1,040,163

APPENDIX II

RED HORSE COMBAT DEFENSE TEAMS

Units of RED HORSE Combat Defense Teams

From 10 August 1968 to 31 May 1969, a total of 17,000 man-hours were directed toward training and deploying RED HORSE CDTs to combat defense situations. Included in this total were weekly firings and orientations by the ROK Infantry stationed at Phan Rang. In addition to the command post and field command post, these units are included in the RED HORSE Combat Defense Teams:

Strike Force - Two radio controlled strike teams are employed. They consist of two jeeps monted with M-60 machine guns and one radio controlled 2 1/2-ton truck mounted with 50-caliber heavy armament machine gun team.

Five 50-man Defense Teams - All members are armed with M-16 rifles. In addition, teams are equipped with M-60 machine guns and M-79 rocket grenade launchers and shotguns. Each squad leader has ample supplies of white slap flares for illumination.

Air Police Augmentees - Twenty-five combat trained airmen, twenty regulars and five reserves.

Armory Team - Issue weapons, slap flares, claymore mines, ammunition, etc.

Demolition Team - One demolition expert is assigned to each defense team and sets explosives, trip flares and claymore mines as directed by the RED HORSE commander.

Disaster Recovery Team - Provides emergency repair resulting from rocket or mortar attack at the request of the 35th TFW commander.

Area Defense Team - Guards the RED HORSE compound, base power plant, water treatment facility, LOX plant, the Wing Commander's and Colonel's quarters area, and provides back-up to other RED HORSE teams.

Dispatch - Assures alert vehicles are ready on line, fueled and ready to roll in case of an emergency. Dispatches vehicles on the command of the Commander or Command Post.

Reserve Team - Team of 20-25 NCOs and airmen armed and fully mobile in a standby alert for dispatch to assist Air Police or to support RED HORSE combat teams requiring additional forces or to replace wounded personnel.

Food Service Team - Provides coffee and rations to troops in the field during sustained attack.

Deployment of Combat Defense Teams

Combat Defense Teams were deployed 14 times during mortar/rocket and infantry/sapper attacks. On numerous other occasions, RED HORSE was called upon to man defense positions because of increased activity by the enemy. These statistics summarize the hostile action necessitating deployment of Combat Defense Teams:

ATTACK	TIME	MORTARS	ROCKETS	VC INFANTRY/SAPPERS	
26 Jan 69	0115-1000	86	8	Yes	7 hrs
1 Feb 69	0030-0330	20+	0	Unk	
2 Feb 69	0100-0330	35+	9	Yes	2 1/2 hrs
11 Feb 69	0200-0430	25+	8	Unk	
12 Feb 69	0141-0445	86	0	Yes	1 hr
24 Feb 69	0225-0600	10	0	0	0
15 Mar 69	0130-0250 / 0550-0750	0	5	0	
16 Mar 69	1930-2145	34	7	0	
19 Mar 69	0245-0745	23	4	0	
21 Mar 69	2255-0110	25	2	0	
23 Mar 69	0155-1510	32	0	Yes	1 1/2 hrs
24 Mar 69	2400-0600	41	5	Unk	

Defense Preparations

Defense preparations of RED HORSE Combat Defense Teams included:

1. Two miles of triple concertina wire have been laid in front of defense positions.

2. Eighty-two sandbag bunkers have been constructed.

3. Three dug-in sand bunkers for the Strike Force have been built.

4. A 12,000 LF zig zag barb wire fence with trip flares attached has been constructed by RED HORSE troops. Over 600 trip flares have been placed on this fence, and in a 50-ft. wide mined area adjacent to the concertina barrier.

5. Defense barrier includes a double apron barb wire fence in those positions most vulnerable for an attack.

6. Approximately 40 acres of brush have been cleared to allow for a free fire zone in front of the defense line.

7. A two-mile dirt access road immediately behind defense bunkers was constructed to facilitate movement of troops, replenishment of ammunition, and field feeding of troops.

8. Three personnel bunkers were constructed in the compound area. Trigger shields were built over all bunkers, as well as the Command Post.

9. An airhorn was designed and installed at the Command Post (activated by a push button inside of the Command Post) to alert the 554th CES personnel of incoming mortar/rockets and/or alert them to report to their predetermined defense teams assemble points.

APPENDIX III
555th CES (HR) PROJECTS

AS OF 26 JUNE 69

PROJECTS COMPLETED SINCE 4 DEC 68

	PROJECT NO	DESCRIPTION	START CONST	CONST COMPL
1.	721 HR	Aircrew Quarters		21 Dec 68
2.	69-7	Rec Workshop	27 May 68	15 Jan 69
3.	98-8	Clothing Sales Store	7 Oct 68	22 Jan 69
4.	106-8	F-4C Utilities		31 Jan 69
5.	244-7	Fire Training Area	17 Apr 68	15 Jan 69
6.	16-8	Lean-To Eng Prop	17 Apr 68	8 Feb 69
7.	147-8	Showers, Latrines	11 Nov 68	12 Feb 69
8.	129-9	Preload Gantry	3 Jan 69	15 Feb 69
9.	143-9	Arm/DeArm Shelter	31 Jan 69	17 Feb 69
10.	219-9	Guard Tower	7 Feb 69	20 Feb 69
11.	205-9	VIP Revetments	17 Feb 69	12 Mar 69
12.	148-8	VASI Lighting	5 Dec 68	15 Mar 69
13.	68-8	Radio Beacon	2 Jan 69	19 Mar 69
14.	178-9	D2, 556 Recon Sq Ops	10 Feb 69	21 Mar 69
15.	172-6	BX Snack Bar	15 Oct 68	7 Apr 69
16.	186-9	Underground Wiring (412 MMS)	21 Mar 69	5 Apr 69
17.	118-8	Ration Breakdown	1 Mar 69	11 Apr 69
18.	202-9	Repair BAK-12	5 Jan 69	12 Apr 69
19.	201-9	Const Revetments C-130	7 Jan 69	12 Apr 69
20.	80-8	Kennel	10 Feb 69	26 Apr 69

PROJECT NO	DESCRIPTION	START CONST	CONST COMPL
21. 953 HR	Market Time Road	1 Apr 69	28 Apr 69
22. 126-9	POL Dikes	10 Feb 69	28 Apr 69
23. 72-8	ACE Facility	1 Mar 69	9 May 69
24. 108-8	Water & Sewer Lines Bldg 615	26 Apr 69	12 May 69
25. 105-9	Dormitory, 28 Man	15 Mar 69	13 May 69
26. 123-8	Install Runway Radar Reflectors	17 Mar 69	14 May 69
27. 125-8	Vehicle Maint Bldg West	4 Dec 68	22 May 69
28. 255-7	Water Mains East	4 Dec 68	21 May 69
29. 149-9	Acnd Bldg 4400 (Vet Clinic)	16 Apr 69	29 May 69
30. 155-9	Alter Serving Lines Dining Hall #5	23 Mar 69	6 Jun 69
31. 152-9	Const AGE Support Fac	31 Mar 69	10 Jun 69
32. 161 HR	Const Roads	20 Nov 68	21 Jun 69
33. 20-8	Const Post Office	3 Apr 69	24 Jun 69

PROJECTS UNDER CONSTRUCTION

PROJECT NO	DESCRIPTION	PRIORITY	SCOPE	EST TOT COST
1. 38-9	BNT BAK-12 Barrier	6	2 ea	28,800
2. 168-9	Repair Plastic Water Lines	25A	4,500LF	6,000
3. 78-8	Beach Road	52	70,000SY	190,000
4. 150-8	Repair Parking Lots	62	60,000SY	90,500
5. 141-8	Const Sidewalks	66	14,600LF	35,700
6. 226-9	Const Rec Fac (GYM)	66A	10,000SF	49,200
7. 171-9	Chapel Modulars	88	3 ea	223,200
8. 230-9	Erect Education Center	92	3 ea	26,200
9. 95-8	Warehouse Supply Bldg.	5	10,000SF	20,000

PROJECT NO	DESCRIPTION	PRIORITY	SCOPE	EST TOT COST
10. 220-7	Const Duc Fac (Telephone)	8A	4,344LF	49,000
11. 169-9	Revetments, F-4C	1A	3,400LF	49,500
12. 258-9	Pave Supply Yards	60	34,000SY	106,500
13. 122-8	Radar Maint Fac (GCA)	51	640SF	7,400
14. 113-9	Repair AMMO Storage Area	30	32 ea	64,300
15. 213-9	Const Maint Docks, C-7A	16	2,482SF	35,100
16. 115-8	Const Fuel Storage Dikes	28	2,285LF	6,100
17. 62-8	Const Security Lights & Fence	56	4,650LF	47,800
18. 96-8	Acnd Red Cross Fac	63	4,000SF	9,900
19. 126-8	Const Vehicle Refuel Stn	41B	2,680SF	47,300
20. 979 HR	Sqdn Ops (68S 087301)	1	4,000SF	44,000
21. 283-9	Const Warehouse	26	10,000SF	39,100
22. 142-9	Const Officers Qtrs East Side		2,880SF	40,300
23. 275-9	Const Acrft Revetments		4,000LF	43,500
24. 278-9	Const Org Maint Bldg		4,800SF	30,100
25. 279-9	Const Gen Purp Acnd Shop (Admin)		4,800SF	32,100
26. 281-9	Const Maint Fac		11,600SF	46,500
27. 300-9	Const Taxiways & Road		7,433SY	39,900

BINH THUY O&M PROJECTS

PROJECT NO	DESCRIPTION	PRIORITY	SCOPE	EST TOT COST
1. 41-8 (R-1)	Const Subsistence Whse	14	4,800SF	14,900
2. 11-9	Const Revet, Passive Def	10A	1,330LF	8,600
3. 19-9 (R-1)	Const Chapel	12	4,460SF	79,400
4. 56-9	Maintain Airfld Pvt (Seal Coat)	12	183,200SY	27,000

PROJECT NO	DESCRIPTION	PRIORITY	SCOPE	EST TOT COST
5. 42-8	Const Acrft Shelters	16	4 ea	,600
6. 46-9	Const Sanitary Latrines	8	840SF	16,100
7. 33-8 (R-1)	Const Canine Kennel		BCE Project	
8. 14-8	Const Base Laundry		Not Approved	

PROJECTS UNDER DESIGN/AWAITING SCHEDULING

PROJECT NO	DESCRIPTION	PRIORITY	SCOPE	EST TOT COST
1. 203-9	Install BAK-13	7B	1 ea	14,100
2. 233-9	Sewage Treatment & Disp RMK	21A	10,000SY	29,600
3. 157-9	Whse Sup * Equip Base	25A	5,000SF	21,400
4. 231-9	Install Fuel Solids Sepr	28A	2 ea	19,700
5. 228-9	Const Comm Center (AUTODIN)	41B	546SF	8,600
6. 177-8	Const TACAN Fac	43	534SF	36,300
7. 227-9	Repair Runway Shoulders	44	55,500SY	22,500
8. 164-9	Acnd Library & Tape Center	65	15 Tons	9,900
9. 140-8	Alter Machine Rm, Bldg 505 A/C	66	3,785SF	2,900
10. 182-9	Revetments, Passive Def	67	4,400LF	29,900
11. 89-8	BAK-13 Corrections	67	1 ea	37,600
12. 101-8	C-7A Water Mains	29B	15,000LF	34,500
13. 173-9	AM-2 Apron	32A	4,200SF	16,300
14. 242-9	Const Revetments, AMMO	14	236LF	6,500
15. 254-9	Const Sewer Lines & Lift Stn	18	2,000LF	20,600
16. 160-8	Const Base Theater	70	8,640SF	44,200
17. 253-9	Const UHF/VHF Fac	13	1,490SF	40,200
18. 102-9	Maintain TriService Road	56	1.5 mi	45,100

PROJECT NO	DESCRIPTION	PRIORITY	SCOPE	EST TOT COST
19. 172-9	Const Parachute Shop	52	5,760SF	34,500
20. 301-9	Repair JP4 Fuel Stor Dikes	31	LS	154,000
21. 256-9	Const Tire Shop (Auto Maint)	43	2,400SF	15,200
22. 238-9	Install Blast Fence	58	250LF	3,800
23. 131-8	Const Security Fence & Lights	23	4,000LF	35,900
24. 156-9	Const Security Lights (MotorP)	24	LS	16,900
25. 147-9	Alter RC Hospital, Bldg 4770	84	3,120SF	26,300
26. 306-9	Const Storage Fac, East Side	66	9,600SF	35,200
27. 207-9	Const Vault (Bank of America)	65	1 ea	9,600
28. 307-9	Const Comm Cable Duct Fac	8B	LS	15,300
29. 305-9	Const Sup & Maint Fac, West	54	3,840SF	41,600
30. 223-9	Maintain Primary Rwy (Erosion)	38	111,000SY	109,500
31. 138-9	Const Stor Area Fuel Tank & Mobile Trl	62	2,222SY	5,700
32. 269-9	Const Latrine Add't (14 AP)	27	800SF	25,800
33. 320-9	Const Acrft Revetments (Navy)	20A	6,540LF	57,200
34. 101-8(R-1)	Develop Water Wells (HTI)	26	2 ea	8,000

TOTAL SCOPE OF WORK ACCOMPLISHED TO DATE

TYPE OF WORK	VOLUME/QUANTITY
Vertical Construction	83,062 SF
Water Lines	20,445 LF
Sewer Lines	1,465 LF
Concrete	5,630 CY
Electric Lines	3,240 LF
Bricks	46,726
Earthwork	370,000 CY
Rock Utilized	88,196.8 T
Asphalt	2,803.1 T
Fence	1,052 LF
ARMCO Revetment	9,890 LF
Runway Matting Laid	35,000 SF
Runway Matting Removed	552,250 SF

BINH THUY EXPANSION

ITEM	SCOPE	TYPE CONST	PCE ($000)
Apron	40,000SY	M8Ao1	80.0
POL Facilities			
Operations Bldg	1,600SF	Wood (20'X80')	7.4
Mogas Station	1 - OL	Add't Pump	1.0
POL Storage	3,000BBL	Steel tank, lines, etc	30.4
Airfield Lighting	3,000LF	Approach lights, beacon and wind indicator	71.2
Operations Facility	3,200SF	Wood (30'X108')	14.9
Air Freight Terminal	2,900SF	Wood (30'X96')	19.7
Open Storage	20,000SY	Asphalt	48.0
Covered Storage(Wshe)	25,000SF	1 - PASCOE (160'X100') 1 - PASCOE (144'X70')	47.6
Wing Headquarters	10,000SF	Wood (30'X108')	46.5
AMN Dorm	40MN	Wood Hootch (20'x144') 10 Bldg	123.0
Latrines, AMN	40MN	Wood (20'X80') 2 Bldg	16.0
BOQ	120MN	Wood Hootch (20'X108') 7 Bldg	49.1
Dining Hall	10,500SF	Wood	51.0
Auto Maint Shop	8,400SF	Wood (40'X108') 2Bldg	46.3
AMMO Maint Shop	10,276SF	Wood (40'X132') 2Bldg	42.4
Ref Maint Shop	2,700SF	Wood (30'X90')	10.8
Utilities:			
Elect	1,000KW	3 350KW or 2 500KW Gen Trans Lines	360.0
Water 1	150TG	Lines, Stor Tank & Booster Pump	240.0
Site Prep	50,000CY	Select Fill	417.0
Roads/Parking Lots	26,800SY	2" Asphalt with 6" base	178.5
		TOTAL	2,149.8

WORK LOAD	NUMBER	TEG	LABOR (49%)	MATERIAL & EQUIP (50%)
Projects completed since 4 Dec 68				
CRB	33	814.7	396.8	417.9
BNT	2	31.8	15.6	16.2
TOTAL	35	846.5	412.4	434.1
Work in Support of other RH units	7	455.6	236.9	N/A
Projects under construction	27	1,408.0	689.9	718.1
Binh Thuy O&M Proj	8	1,466.0	718.3	747.7
Binh Thuy Expansion	22	2,149.8	1,160.9	988.9
Proj Under Design/ Awaiting Scheduling	34	1,036.0	507.6	528.4
TOTAL SCOPE	126	6,906.3	3,489.1	3,417.2

APPENDIX IV

820th CES CONSTRUCTION PROGRAM

		NO	TOTAL COST (000)
I.	UNDER CONSTRUCTION	44	3,583.7
II.	APPROVED/DESIGN	74	2,959.2
III.	PENDING 7AF APPROVAL	99	3,136.2
	TOTALS:	<u>217</u>	<u>9,679.1</u>

1 MAY 1968 TO 1 MARCH 1969

COMPLETED PROJECTS DA NANG

	NAME	NO.	SCOPE	TOT. COST(000)
A.	STRUCTURES	64	465,558 SF	1,841.7
B.	AIR FIELDS/ROADS	3	37,210 SY	80.8
C.	UTILITIES/POL	8	N/A	186.8
D.	REVETMENTS	2	18,270 LF	2,606.7
	TOTALS:	<u>77</u>		<u>4,716.1</u>

UNDER CONSTRUCTION AFTER 31 MAY 1969

	NAME	NO.	SCOPE	TOTAL COST (000)
A.	STRUCTURES	15	73,658 SF	882.3
B.	AIRFIELDS/ROADS	2	106,755 SY	385.0
C.	UTILITIES/POL	5	N/A	171.6
D.	REVETMENTS	4	12,550 LF	240.7
	TOTALS:	26		1,679.6

EMERGENCY BOMB DAMAGE REPAIR

ASP-1 BOMB STORAGE AREA

BOMB STORAGE AREA	% COMP	EST COMP	EST COST
738-9 Pre-Load Facility	80	17 Jun 69	30.0
740-9 MMS Admin Bldg	100	8 May 69	15.0
741-9 Missile Maint	100	23 May 69	15.0
747-9 Bomb Stor Bldg (3)	15		95.0
750-9 Rocket Assem Bldg	95	15 Jun 69	10.0
742-9 Stor Sheds (5 ea)	0	15 Jul 69	35.0
751-9 Missile Maint Stor	100	26 May 69	10.0

TOTAL MAN-HOURS EXPENDED TO DATE: 20,234

TOTAL COST TO DATE: 52,008.63

AIRCRAFT SHELTER PROGRAM

RED HORSE ERECTION

PHASE I	55
PHASE II (OLD AERIAL PORT AREA)	19
PHASE III (VNAF)	18
PHASE IV (F-102 AREA)	6
TOTAL:	__98__

RED HORSE COVER

COVER COMPLETE	20
COVER IN PROGRESS	25
EQUIVALENTS	27

CONTRACTOR COVER

COVER COMPLETE	35
COVER IN PROGRESS	0

RED HORSE COST TO DATE: 1,930,000.00

APPROVED/UNDER DESIGN

	NAME	NO.	SCOPE	TOTAL COST (000)
A.	STRUCTURES	23	109,201 SF	1,215.3
B.	AIRFIELDS/ROADS	12	33,354 SY	333.1
C.	UTILITIES/POL	5	N/A	107.0
D.	REVETMENTS	2	640 LF	9.8
	TOTALS:	__42__		__1,665.2__

PROGRAMMED: PENDING 7AF APPROVAL

	NAME	NO.	SCOPE	TOTAL COST (000)
A.	STRUCTURES	35	244,537 SF	1,418.1
B.	AIRFIELDS/ROADS	22	182,679 SY	931.8
C.	UTILITIES/POL	5	N/A	133.3
D.	REVETMENTS	1	640 LF	24.9
	TOTALS:	63		2,508.1

UNDER CONSTRUCTION (TUY HOA)

	NAME	NO.	SCOPE	TOTAL COST (000)
A.	STRUCTURES	13	80,048 SF	995.0
B.	AIRFIELDS/ROADS	4	293,191 SY	909.1
C.	UTILITIES/POL	-	-	-
D.	REVETMENTS	1	N/A	31.8
	TOTALS:	18		1904.1

APPROVED/UNDER DESIGN (TUY HOA)

	NAME	NO.	SCOPE	TOTAL COST (000)
A.	STRUCTURES	17	32,340 SF	544.3
B.	AIRFIELDS/ROADS	6	539,980 SY	544.3
C.	UTILITIES	6	N/A	127.7
D.	REVETMENTS	3	N/A	77.7
	TOTALS:	32		1,294.0

PROGRAMMED: PENDING 7AF APPROVAL

	NAME	NO.	SCOPE	TOTAL COST (000)
A.	STRUCTURES	22	36,165 SF	365.0
B.	AIRFIELDS/ROADS	8	187,974 SY	198.6
C.	UTILITIES/POL	2	N/A	10.2
D.	REVETMENTS	4	N/A	54.3
	TOTALS:	36		628.1

APPENDIX V
PERSONNEL STRENGTH

556th CES (HR) PROJECT SUMMARY

		1May68	30Jun68	1Aug68	15Sep68	20Sep68	15Oct68	20Dec68	30Apr69
UDORN	Officers	1	1	1	1	1	1	1	1
	Airmen	31	32	32	41	42	51	51	43
	WAE (LN)	505	505	383	402	420	450	450	450
TAKHLI	Officers	2	2	2	1	1	1	1	1
	Airmen	24	28	28	16	16	16	15	13
	WAE (LN)	350	350	235	150	150	150	150	150
NKP	Officers	4	3	3	3	3	5	5	4
	Airmen	141	133	133	154	154	170	170	149
	WAE (LN)	1035	1035	912	978	1013	1159	1200	1200
	UMD (LN)	1	1	1	1	1	1	1	1
UBON	Officers	0	1	1	1	1	0	0	0
	Airmen	19	25	25	25	25	13	11	9
	WAE (LN)	260	260	229	229	229	45	151	151
KORAT	Officers	1	2	2	2	2	1	1	1
	Airmen	21	29	29	15	15	14	10	8
	WAE (LN)	240	240	202	202	132	132	150	150
DON MUANG	Officers	1	1	1	1	2	1	1	2
	Airmen	5	6	6	5	5	5	5	5
	WAE (LN)	3	2	2	2	2	2	2	2
	UMD (LN)	10	11	11	11	11	11	11	11
U-TAPAO	Officers	5	6	6	7	7	7	7	7
	Airmen	147	131	131	115	118	122	122	98
	WAE (LN)	630	670	560	560	577	565	420	420
	UMD (LN)	40	38	38	38	38	38	38	38
TOTAL	Officers	14	16	16	16	16	16	16	16
	Airmen	388	384	384	371	391	386	384	325
	WAE (LN)	3023	3062	2523	2523	2503	2523	2523	2523
	UMD (LN)	51	50	50	50	50	50	50	50

NOTE: Officers and airmen shown as assigned strength; WAE & UMD shown as authorized. Assigned strength fluctuated, but overall sqn LN strength stayed within 10% of authorized.

556th CES WORK UNITS ACCOMPLISHED

Land Cleared	4,063	Acres
Roads Paved	20.3	Miles
Roads Constructed	19.2	Miles
Airfield Pavements Paved		
Runway	194,295	Syds
Taxiway	175,707	Syds
Airfield Shoulder Stabilized	70,859	Syds
Aircraft Revetments Constructed	58,090	SF
Airfield Matting Installed	43,000	Syds
Airfield Matting Removed	263,767	Syds
Buildings Constructed		
Laundry	24,200	SF
Warehouses	112,641	SF
Shops/Hangars	81,283	SF
Recreation Facilities	47,860	SF
Administration Facilities	52,503	SF
Operational Facilities	272,317	SF
Sidewalks	14,960	Syds
Dining Halls	18,600	SF
Parking Facilities	51,668	Syds
Dormitories (Incl. Latrines)	144,315	SF
Barriers (BAK-12+13)	6	EA
Pre-Engineered Metal Bldgs	58	EA
Hospitals, Modular	4	EA

Utilities Constructed

VASI	5	EA
Water Lines	55,470	LF
Sewage Lines	44,126	LF
Sewage Lagoons	20.6	Acres
Power Lines	127,800	LF
Water Treatment	720,000	Gal/Day
Septic Tanks	115,000	Gal
Ducts & Conduits	2,575	LF

	U-TAPAO			
Project Description	CWE	Actual Proj Cost	Const Start Date	BOD
67SMCP 211-193 Engine Test Stand	95.0	88.0	Sep 67	Jun 68
67SMCP 214-425 Refueling Veh Maint, 14,400SF	125.0	119.7	Jul 67	Jun 68
P458 60-7 Base Procurement, 4000SF	22.2	21.8	Aug 67	Aug 68
P458 71-7 SAC Antenna Farm	25.0	24.5	Jan 68	Aug 68
P458 79-7 Veh Refueling Stor Open	23.5	4.5	Mar 68	Aug 68
67SMCP 218-712 GSE West, 9000SF	31.5	26.4	Nov 67	Aug 68
67SMCP 510-001 Composite Medical Fac	164.0	165.5	Nov 67	Aug 68
68SMCP 422-25A Bomb Preload, 4 ea	3.0	2.2	Jul 68	Sep 68
P458 111-7 Const Open Storage 7000SF	2.0	2.4	Apr 68	Sep 68
P458 117-7 Education Center 4800SF	25.0	19.7	Dec 67	Aug 68
P458 33-9 MMS Parking Area	8.8	5.2	Aug 68	Sep 68
P458 6017-9 Bomb Damage Apr	35.0	40.7	Aug 68	Sep 68
P458 85-8 Air Cond POL Ops Lab	0.6	0.3	Oct 68	Oct 68
67SMCP 730-841 Dog Kennel	50.0	45.7	Aug 68	Oct 68
P458 89-8 TAFDS, POL Dikes, C Ramp	5.3	0.6	Aug 68	Sep 68
66SMCP 812-223 OIH Distrib	169.0	164.3	Jun 67	Nov 68
67SMCP 952-586 Revetments (70% compl)	169.0	116.4	Nov 67	Sep 68

U-TAPAO

Project Description	CWE	Actual Proj Cost	Const Start Date	BOD
67SMCP 713-366 Relocate Trlr. Ct	121.9	88.6	Nov 67	Dec 68
68SMCP 722-122R Dorm Modular, 880MN	120.0	104.7	Feb 68	Dec 68
68SMCP 932-586 15 Revetments	205.0	41.0	Apr 68	Dec 68
P458 162-7 LSE Fac 4800SF	40.2	19.2	Mar 67	Dec 68
P458 1-7 20,000LF Security Fence	10.0	9.9	Feb 68	Jan 69
P458 86-8 Prime Elec Distrib, OH	16.2	15.3	Mar 68	Jan 69
P458 46-9 POL Dikes, TAFDS	2.9	2.9	Nov 68	Jan 69
P458 75-7 Bx Warehouse	41.8	14.7	Jan 69	Feb 69
P458 85-7 POL Storage 4800SF	18.1	17.1	May 68	Feb 69
P458 132-7 Reproduction Center	30.0	20.4	May 68	Feb 69
P458 22-9 ACND PDO	4.8	2.0	Jul 68	Feb 69
P458 5744-9 Air Rescue Facility	10.0	9.0	Feb 69	Feb 69
P458 51-7 Transceiver Bldg	25.0	24.9	Jul 68	Jan 69
*67SMCP 116-945 Blast Deflectors, 2000LF	50.0	40.0	Dec 67	Mar 69
*67SMCP 135-666 VASI System	8.0	8.0	Feb 68	Mar 69
*67SMCP 730-711 Base Landry 14,400SF	75.0	60.0	Aug 67	Apr 69
*67-68SMCP 740-618 Officers Club Addn 15,500SF	371.0	300.0	Sep 67	Mar 69

U-TAPAO

Project Description	CWE	Actual Proj Cost	Const Start Date	BOD
*67SMCP 800-000 Water & Sewage	10.0	10.0	Aug 68	Mar 69
*68SMCP 300MN Modular Dinning Hall	70.0	33.0	May 68	Feb 69
*68SMCP 442-25AR Revet Trlr Holding	25.5	10.0	Jun 68	Mar 69
*P458 83-7 R&M Whse	32.1	33.0	Aug 68	Apr 69
*P458 166-7 Clothing Sales	24.0	22.0	Dec 68	Mar 69
		1,733.6		
WORK PLACED, NOT BOD'd		252.0		
TOTAL		1,985.6		

Project Description	UDORN CWE	Actual Proj Cost	Const Start Date	BOD
P458 12-9 Comm & Elec Sh Addn	1.7	1.5	Dec 68	Jan 69
P458 55-9 Erect Sh Acft Maint, 2BUSH	11.6	11.6	Dec 68	Jan 69
P458 97-9 Const Shop Acft Maint	3.8	3.5	Dec 68	Jan 69
67SMCP 136-666 VASI	10.0	9.3	Aub 68	Mar 69
P458 47-8 Auto Admin Fac	23.5	23.0	Dec 68	Feb 69
P458 63-8 CRC Addn	23.0	22.9	Jul 68	Feb 69
P458 71-8 Shoulder, Runway Access	20.1	20.1	Jan 69	Feb 69
P458 181-8 Alter BOQ	9.6	9.6	Dec 68	Feb 69
P458 28-9 RPS Amn Dining Hall	1.2	1.0	Jan 69	Feb 69
P458 104-9 Maintain Apron Shoulder	16.0	16.0	Dec 68	Feb 69
*66SMCP 932-586 Revetments, 14,620LF	166.0	139.0	Apr 68	Apr 69
*P458 59-8 Refueler Parking	23.3	22.9	Dec 68	Mar 69
*P458 65-8 Main Gate	9.1	9.1	Sep 68	Nov 68
*P458 167-8 Av Oil Storage	12.5	8.8	Nov 68	Apr 69
*P458 178-8 Al Photo Lab, Recon.	7.6	7.2	Jan 69	Apr 69
*P458 191-8 Alter Laser Shop	9.9	8.8	Nov 68	Mar 69
*P458 78-9 Oper Apron	24.5	22.1	Dec 68	Apr 69

Project Description	UDORN CWE	Actual Proj Cost	Const Start Date	BOD
P458 70-8 Autodin Addn Supply	6.6	6.6	Jul 68	Oct 68
P458 202-8 Alter Ops Spce, 7AACS	6.1	6.2	Sep 68	Oct 68
P458 65-8 Const Main Gate	7.5	7.2	Sep 68	Nov 68
P458 96-8 Shelter WX Instr	2.5	2.5	Sep 68	Nov 68
P458 114-8 Const Latrine	2.0	2.0	Sep 68	Nov 68
P458 199-8 Addn to Air Freight	4.2	4.2	Nov 68	Nov 68
P458 54-9 Const Orderly Room	2.6	2.6	Oct 68	Nov 68
67SMCP 510-001R 50 Bed Modular Hosp	80.0	76.7	Feb 68	Dec 68
P458 4-7 Para Dinghy Bldg	23.8	23.8	May 68	Dec 68
P458 206-7 Library Education	20.1	20.1	Sep 68	Dec 68
P458 135-8 CE Covered Storage	16.5	16.5	Oct 68	Dec 68
P458 172-8 Alt. A/C Gen Purp Shop	11.8	11.8	Oct 68	Dec 68
P458 80-9 Const Latrine, RAPCON	1.0	1.0	Dec 68	Dec 68
P458 118-9 Apron, Service Vehicle	2.9	3.0	Dec 68	Dec 68
P458 71-8 OSE Shop Addn, 1800SF	9.1	8.9	Dec 68	Jan 69
P458 200-8 Const Latrine	4.2	3.9	Dec 68	Jan 69
P458 200-8 Const 7 Accs Maint Fac	11.4	11.4	Dec 68	Jan 69

Project Description	UDORN CWE	Actual Proj Cost	Const Start Date	BOD
P458 36-8 Airfield Matting	24.4	21.4	Jan 68	May 68
P458 73-8 Intelligence Addition	5.0	4.9	Apr 68	May 68
P458 35-8 Addn to Bx Warehouse	10.0	8.8	Mar 68	May 68
P458 81-8 RTAF CE Bldg	10.0	9.0	Apr 68	Jun 68
P458 83-8 Addn Officer Billet 314	10.0	9.1	Apr 68	Jun 68
P458 85-8 POL Farm Parking	6.7	6.6	Jun 68	Jul 68
P458 117-8 Const Maint Supply	5.5	5.5	Jun 68	Jul 68
P458 7-8 Const Latrines	17.8	16.5	Jun 68	Aug 68
P458 11-8 Const Personnel Shelter	9.9	9.9	Jun 68	Aug 68
P458 64-8 const Slab Port Snack Bar	3.0	2.8	Aug 68	Aug 68
P458 102-8 Expand Mainl Room	4.9	4.7	Jul 68	Aug 68
P458 174-7 Inflight Kitchen	14.9	10.4	Jun 68	Sep 68
P458 43-8 BASO Admin Addn	19.9	15.0	Apr 68	Sep 68
P458 7-9 Install BAK-13	24.5	9.9	Aug 68	Sep 68
P458 19-9 Computer Rm, BASO	15.6	8.0	Aug 68	Sep 68
P458 71-9 Shop, Aircraft Maint	1.4	1.0	Sep 68	Sep 68
P458 218-7 Subsistance Storage	24.2	13.4	Aug 68	Oct 68

UDORN

Project Description	CWE	Actual Proj Cost	Const Start Date	BOD
P458 105-7 Auto Maint Shop	37.5	23.0	Jan 69	Apr 69
P458 131-7 Lanundry	48.0	48.0	Jan 69	Apr 69
P458 77-8 TV Studio	24.9	23.0	Jan 69	Apr 69
P458 116-8 Const Sentry Towers	24.9	24.0	Jan 69	Apr 69
P458 137-8 Install Lights, Supply	8.0	8.0	Jan 69	Apr 69
P458 185-8 Combat Ops Center	24.6	<u>23.0</u>	Jan 69	Apr 69
TOTAL		850.7		

TAKHLI

Project Description	CWE	Actual Proj Cost	Const Start Date	BOD
P458 70-7 CBPO Concrete Block	25.0	25.0	Sep 67	Jun 68
67SMCP 142R 25 Bed Modular Disp	39.0	29.1	Jan 67	Sep 68
P458 139-8 Spcl Flt Ops	19.0	18.0	Apr 68	Sep 68
P458 146-8 Maint Debrief 960SF	9.6	6.9	Mar 68	Aug 68
P458 253-7 Intelligence Facility 5600SF	7.8	7.9	Feb 68	Oct 68
66SMCP 932-586 Aircraft Revetments	120.0	104.4	Oct 67	Oct 68
P458 110-7 Elect Dist System	24.5	17.2	Apr 67	Oct 68
P458 124-7 Med Supply Prefab	24.1	22.9	Aug 67	Oct 68
P458 252-7 Whse Supply & Equip	38.7	33.5	Mar 68	Oct 68
P458 9-7 Clothing Sales/Laundry	20.7	13.4	Jun 67	Oct 68
P458 143-8 Seg Ammo Storage	9.6	7.5	May 68	Sep 68
P458 251-7 Subsistance Stor Whse	15.6	7.8	Aug 68	Nov 68
66SMCP 722-211 Dorm, 256 MN	137.9	72.3	Sep 68	Jan 69
P458 161-8 Const Shelter AGE	10.0	6.7	Dec 68	Feb 69
P458 179-8 Flammable Storage	4.3	4.1	Jan 69	Feb 69

TAKHLI

Project Description	CWE	Actual Proj Cost	Const Start Date	BOD
P458 83-8 AGE Serv Shelter	22.0	19.1	Jan 69	Feb 69
P458 1-9 Gun Cleaning Pads	23.5	18.0	Jan 69	Feb 69
TOTAL		413.8		

Project Description	NKP CWE	Actual Proj Cost	Const Start Date	BOD
P458 75-8 Airmen Dorm 80 MN	27.5	26.6	May 68	Oct 68
66SMCP 112-213 Taxiway, Runway Access	173.0	206.3	Nov 67	Nov 68
66SMCP 216-642 Shop, Ammo Maint 2370SF	24.0	36.4	Dec 67	Nov 68
66SMCP 610-243 Hq Air Base Gp 14,000SF	73.0	71.4	Jan 68	Nov 68
67SMCP 740-618 Officers Open Mess 7400SF	98.0	93.1	Sep 67	Nov 68
P458 145-7 Whse Subsistance 4000SF	11.4	12.6	May 68	Nov 68
P458 70-9 Repair Taxiway 18,900SY	196.5	166.3	Oct 68	Nov 68
P458 75-9 Site Clearing	5.6	2.8	Sep 68	Nov 68
67SMCP 740-617 NCO Open Mess Addn	120.0	118.7	Sep 67	Dec 68
67SMCP 113-322 Apron, Hangar Access	25.0	11.0	Sep 68	Dec 68
67SMCP 116-622 Pad, Grnd Sta & Road	12.0	7.1	Sep 68	Dec 68
67SMCP 211-112 Hangar, Org Maint 10,000SF	90.0	88.1	Feb 68	Oct 68
67SMCP 211-111 Hangar Maint & Shop	108.0	80.2	Sep 68	Dec 68
67SMCP 800-000 Utilities Addn	50.0	30.0	Aug 67	Dec 68
P458 72-8 Pacer Dog Radar Pad	14.3	1.7	May 68	Dec 68
P458 67-9 Emer Rnwy Rpr	124.0	117.5	Aug 68	Dec 68
66SMCP 112-215 Taxiway, Apron Access 87,840SY	878.0	878.0	Sep 66	Jan 69

Project Description	NKP CWE	Actual Proj Cost	Const Start Date	BOD
P458 MUK 4-9 Cantonments	20.4	19.8	Dec 68	Jan 69
P458 MUK 5-9 Operations Fac	20.4	18.1	Dec 68	Jan 69
*66SMCP 851-147 Roads, 10.2 Mi	819.0	810.0	Jul 68	Mar 69
*67SMCP 111-118 Rehab Rnwy, 8000FT	1,375.0	1,370.0	Nov 68	Apr 69
*67SMCP 116-922 BAK-12 Barrier	49.0	49.0	Dec 68	Apr 69
*67SMCP 136-666 VASI	13.0	13.0	Jan 68	Apr 69
*67SMCP 452-252 Base Open Storage	70.0	65.0	May 68	Apr 69
*67SMCP 722-211 60MN BOQ	91.0	55.0	Dec 68	Mar 69
*67SMCP 740-676 Rec Multi Purpose	143.0	142.0	Feb 68	Apr 69
*67SMCP 932-586 TFA Revetment	60.0	62.0	Nov 68	Mar 69
*P458 91-6 TACAN	24.8	22.0	Jul 68	Jan 69
*P458 8-7 Rnwy Dist Markers	25.0	25.0	May 68	Apr 69
*P458 112-7 Pass/Air Freight Term	26.0	24.0	Oct 68	Apr 69
*P458 176-7 Squad ops	26.0	24.0	Feb 68	Apr 69
*P458 195-7 CRP Ops	10.0	10.0	Jan 69	Apr 69
*P458 198-7 Laundry	20.0	19.0	Apr 68	Apr 69
*P458 52-8 Comm Transp	14.1	11.0	Jun 68	Mar 69

Project Description	NKP CWE	Actual Proj Cost	Const Start Date	BOD
67SMCP 422-251 Storage, Base Ammo	157.1	134.5	Aug 67	Jun 68
67SMCP 113-314 Apron, Operational, 85,574SY	968.0	818.3	Nov 67	Jul 68
67SMCP 730-841 Kennel, Canine	50.0	49.2	Mar 68	Jul 68
67SMCP 141-753 Sqd Ops Addn, 2800SF	25.0	24.6	Feb 68	Jul 68
67SMCP 442-000 Warehouse, Supply & Equip	70.0	69.1	Sep 67	Jul 68
P458 75-7 Exchange Warehouse 3900SF	40.6	25.3	Mar 68	Jul 68
P458 91-7 Sqd Ops 4000SF	37.1	36.0	Nov 67	Jul 68
P458 124-7 Refueling Veh Open Stor	31.7	24.7	Mar 68	Jul 68
P458 66-8 Whse, Supply & Equip	23.1	26.2	Feb 68	Jul 68
66SMCP 422-251 Mapalm Mix & Storage	48.0	37.4	Nov 67	Jul 68
67SMCP 218-852 Para Dinghy Shop, 5800SF	88.0	83.9	Dec 67	Aug 68
67SMCP 116-672 Corrosion Control, 803SY	54.0	51.7	Nov 67	Aug 68
67SMCP 134-373 GCA Turntable	53.0	50.0	Jan 68	Sep 68
P458 2-7 Rocket Assembly Bldg	24.3	25.0	May 68	Sep 68
P458 104-7 Open Storage 6,000SY	24.8	23.3	Oct 67	Sep 68
67SMCP 211-112 Hanger, Org Maint 10,000SF	90.0	87.3	Feb 68	Oct 68
P458 29-7 Data Processing	25.0	24.8	Mar 68	Aug 68

NKP

Project Description	CWE	Actual Proj Cost	Const Start Date	BOD
*P458 67-8 Storage LSE	7.1	7.1	Dec 68	Feb 69
*P458 63-9 Erosion Control Maint	41.0	40.0	Apr 68	Apr 69
P458 65-9 Rpr Off Club Roof	5.3	3.4	Dec 68	Feb 69
*P458 71-9 Sup Admin & Tool Crib	9.7	9.1	Feb 69	Apr 69
*P458 72-9 Whse S&E, Elec	10.2	10.0	Feb 69	Apr 69
*P458 73-9 Reloc & Enlg Stor	10.2	10.0	Feb 69	Apr 69
*P458 74-9 Const Open Stor	8.1	8.0	Feb 69	Apr 69
*P458 76-9 Veh Maint Shop	10.4	10.4	Dec 68	Mar 69
*P458 77-9 Veh Parking	6.6	6.6	Dec 68	Apr 69
*P458 81-9 Concrete Batch Plant	5.7	5.5	Feb 69	Apr 69
P458 82-9 A/Cond Amn Dorm	13.5	11.3	Jan 69	Feb 69
66SMCP 722-211 Airman Dorm, 216 MN	142.0	135.5	Jun 68	Mar 69
*66SMCP 831-266 Sanitary Sewage Trmnt	79.8	79.8	Mar 68	Apr 69
		6,614.7		
WORK PLACED, NOT BOD'd		1,053.0		
TOTAL		7,667.7		

Project Description	UBON CWE	Actual Proj Cost	Const Start Date	BOD
P458 152-6 Transceiver Bldg	24.9	15.2	Mar 68	Jun 68
P458 71-7 GEEIA Support, Conduits Under Roads	5.1	5.9	Mar 68	Jun 68
67SMCP 136-666 VASI Lighting System	10.9	5.3	Mar 68	Jun 68
P458 116-7 Prefab Subsistence Whse	9.4	11.9	Feb 68	Jul 68
67SMCP 520-142 Modular Dispensary	39.0	39.7	Mar 68	Sep 68
67SMCP 932-586 Aircraft Revetments	40.4	23.5	Mar 68	Sep 68
P458 27-7 Base Roads, 7000SY	26.7	21.7	Apr 68	Aug 68
P458 79-7 Addn Bldg 1320	16.1	25.3	Jun 68	Sep 68
P458 77-7 Addn Bldg 1310 3600SF	26.6	22.8	Jun 68	Oct 68
P458 87-7 Prefab, 2 AGE Shops	24.9	21.9	Feb 68	Oct 68
P458 114-7 Prefab, Pub Dist 6000SF	20.6	18.7	May 68	Oct 68
P458 126-7 Prefab, Med Stor 5100SF	8.6	15.8	Mar 68	Oct 68
P458 190-7 Comm Tandem Switch	15.7	15.8	Jun 68	Nov 68
P458 78-9 BAK-13 Barrier	22.0	11.4	Sep 68	Oct 68
P458 90-7 Veh Maint & Refueling, 2 ea	10.9	14.6	Aug 68	Jan 69
P458 90-7 2 Prefab-Veh & Refuel Maint	20.9	20.9	Aug 68	Jan 69

UBON

Project Description	CWE	Actual Proj Cost	Const Start Date	BOD
P458 120-7 Rnwy Sh Stabilize 26, 600SY	89.4	88.8	May 68	Jan 69
P458 61-7 Airfreight Pack/Crating	11.4	19.1	Dec 68	Feb 69
P458 127-7 Target Intel Fac	10.4	<u>10.4</u>	Jan 69	Apr 69
		408.7		
WORK PLACED, NOT BOD'd		<u>103.0</u>		
TOTAL		511.7		

KORAT

Project Description	CWE	Actual Proj Cost	Const Start Date	BOD
P458 7-8 Comm Tandem Switch	27.0	24.2	Feb 68	Jul 68
67SMCP 932-586 Revetment, Pass Def	47.8	47.8	Feb 67	Aug 68
66SMCP 851-147 Ammo Road	212.0	210.0	Feb 68	Sep 68
67SMCP 136-666 VASI Lighting Sys	10.0	9.8	Dec 67	Oct 68
P458 77-7 Hobby Shop	34.2	33.3	Jun 68	Sep 68
P458 129-6 Base Publications	29.7	22.7	Jul 68	Oct 68
P458 112-8 Bomb Damage, Ammo Area	50.0	24.0	Apr 68	Jul 68
P458 3-8 Tank & Pylon Shop	20.0	19.8	Sep 68	Nov 68
P458 2-9 BAK-13 Barrier	22.0	18.1	Oct 68	Oct 68
P458 84-8 Base Sidewalks	23.5	23.5	Aug 68	Dec 68
P458 19-9 Ammo Haul Road	14.0	12.6	Sep 68	Dec 68
P458 2-8 Veh Maint Bldg	50.0	49.0	Jan 69	Apr 69
P458 2-9 Relocate BAK-13	13.0	13.0	Feb 69	Mar 69
P458 10-9 Erect Revetment	24.6	<u>23.2</u>	Dec 68	Apr 69
		531.0		
WORK PLACED, NOT BOD'd		<u>110.0</u>		
TOTAL		641.0		

APPENDIX VI

SUMMARY OF PROJECT CONCRETE SKY

DIMENSIONS OF SHELTERS:

Most shelters constructed were 68' to 72' in length. Exceptions:

Da Nang	18 ea @ 50'	VNAF A-37s
	4 ea @ 78'	Shelter Curtins
Tan Son Nhut	4 ea @ 76'	RF-101 Acft
	20 ea @ 80'	RF-101 Acft

<u>PROGRAM SCOPE:</u>

1. Number of Shelters Programmed:

Da Nang	98	Phan Rang	61	
Bien Hoa	75	Tuy Hoa	56	
Tan Son Nhut	62	Phu Cat	56	(40)

2. Support Items:

Tan Son Nhut	5 ea dispersal pads for 20 Acft
Phan Rang	9 ea dispersal pads for 45 Acft
Thuy Hoa	2 ea dispersal aprons for 21 Acft. Replacement of 70,000 SY of AM-2 Matting with concrete pavement.
Phu Cat	Replacement of 100,000 SY of AM-2 Matting with concrete pavement.

3. Work Forces:

   a. Pavements At Phu Cat, Phan Rang, and Tuy Hoa constructed by RED HORSE. Dispersal pads at Tan Son Nhut constructed by RMK.

   b. Erection at all bases accomplished by RED HORSE with support of base augmentees. VNAF furnished augmentees for 12 VNAF shelters at Da Nang.

   c. Concrete Cover. RMK covered 53 shelters at Da Nang, 68 at Bien Hoa, and 62 at Tan Son Nhut for a total of 183 shelters. RED Horse is covering 225 shelters, including all bases except Tan Son Nhut.

4. Troop Labor Expenditure:

   Erection: 800 Man hours
   Cover: 1100 Man hours

COST DATA:

1. Total Funded Cost:

Purchase of Shelters	$4,022,800
RED HORSE Erection and Cover Plus Pavements at Phan Rang, Tuy Hoa*	$3,198,000
Contract Cover Plus Dispersal Pads at Tan Son Nhut	$5,414.000
	$12,634,800

2. Average Erection Costs per Shelter:

Funded	$ 300
Unfunded	
Military Labor	$2,400
Equipment Depreciation	140
Overhead (50%)	$1,470
Total:	$4,310

* Cost of Pavements Phu Cat covered under separate project.

3. Average Cover Costs per Shelter:
    a. RMK $26,000
    b. RED HORSE

        Funded $ 9,460 *

        Unfunded

            Military Labor $ 2,940

            Equipment Depreciation $ 1,130

            Overhead (50%)   <u>$ 6,770</u>

    Total $20,300

* Includes $640/Shelter for purchase of Squeez-Crete.

GLOSSARY

A&E	Armaments and Electronics
AB	Air Base
AFLC	Air Force Logistics Command
AFRCE-THAI	Air Force Regional Civil Engineer-Thailand
AFRCE-RVN	Air Force Regional Civil Engineer-Republic of Vietnam
ALCE	Airlift Control Element
AMS	Avionics Maintenance Squadron
ARVN	Army of Republic of Vietnam
ATC	Air Training Command
BAK	Aircraft Arresting Facility
BASO	Base Accountable Supply Office(r)
BCE	Base Civil Engineering
BEMAR	Base Essential Repair Program
BOD	Beneficial Occupancy Date
BOM	Bill of Materials
CEG	Civil Engineering Group
CES	Civil Engineering Squadron
CONUS	Continental United States
CPM	Critical Path Method
EOD	Explosive Ordnance Disposal
FUB	Facilities Utilization Board
HR	Heavy and Repair
IG	Inspector General
LF	Linear Foot
MAC-DC	Military Assistance Command-Directorate of Construction
MCP	Military Construction Program
MECAP	Medical Civic Action Program
MERP	Minimum Essential Repair Road
NCO	Noncommissioned Officer
NKP	Nakhon Phanom
O&M	Operations and Maintenance
OICC	Officer in Charge of Construction
OJT	On-the-Job Training
OMS	Organizational Maintenance Squadron

RED HORSE	Rapid Engineer Deployable Heavy Operations Repair Squadrons, Engineer
ROK	Republic of Korea
RSU	Runway Surveillance Unit
RVN	Republic of Vietnam
SEA	Southeast Asia
SF	Square Feet
SY	Square Yard
TAC	Tactical Air Command
TACAN	Tactical Air Navigation
TASG	Tactical Air Support Group
TDY	Temporary Duty
TFW	Tactical Fighter Wing
TN	Ton

www.ingramcontent.com/pod-product-compliance
Lightning Source LLC
Chambersburg PA
CBHW080550170426
43195CB00016B/2743